Everyday Math Demystified

Demystified Series

Advanced Statistics Demystified
Algebra Demystified
Anatomy Demystified
Astronomy Demystified
Biology Demystified
Business Statistics Demystified
Calculus Demystified
Chemistry Demystified
College Algebra Demystified
Earth Science Demystified
Everyday Math Demystified
Geometry Demystified
Physics Demystified
Physiology Demystified
Pre-Algebra Demystified
Project Management Demystified
Statistics Demystified
Trigonometry Demystified

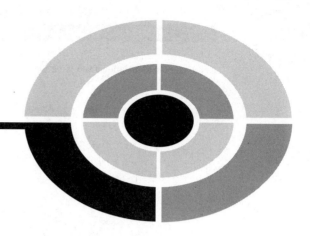

Everyday Math Demystified

STAN GIBILISCO

McGRAW-HILL

New York   Chicago   San Francisco   Lisbon   London
Madrid   Mexico City   Milan   New Delhi   San Juan
Seoul   Singapore   Sydney   Toronto

The McGraw·Hill Companies

Cataloging-in-Publication Data is on file with the Library of Congress

4 5 6 7 8 9 0   DOC/DOC   0 1 0 9 8 7 6 5

ISBN   0-07-143119-5

The sponsoring editor for this book was Judy Bass and the production supervisor was Pamela A. Pelton. It was set in Times Roman by Keyword Publishing Services, Ltd. The art director for the cover was Margaret Webster-Shapiro; the cover designer was Handel Low.

Printed and bound by RR Donnelley.

 This book is printed on recycled, acid-free paper containing a
minimum of 50% recycled, de-inked fiber.

McGraw-Hill books are available at special quantity discounts to use as premiums and sales promotions, or for use in corporate training programs. For more information, please write to the Director of Special Sales, McGraw-Hill Professional, Two Penn Plaza, New York, NY 10121-2298. Or contact your local bookstore.

To Samuel, Tim, and Tony from Uncle Stan

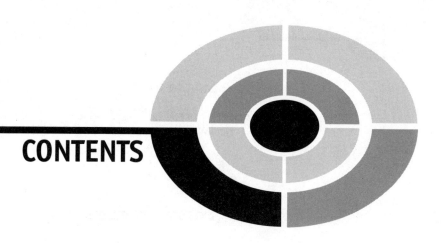

CONTENTS

	Preface	xv
PART 1:	**EXPRESSING QUANTITIES**	**1**
CHAPTER 1	**Numbers and Arithmetic**	**3**
	Sets	6
	Numbering Systems	8
	Integers	10
	Rational, Irrational, and Real Numbers	14
	Number Operations	19
	More Principles Worth Memorizing	23
	Still More Principles	26
	Quiz	30
CHAPTER 2	**How Variables Relate**	**33**
	This versus That	33

Simple Graphs 38

Tweaks, Trends, and Correlation 43

Quiz 49

CHAPTER 3 **Extreme Numbers** **52**

Subscripts and Superscripts 52

Power-of-10 Notation 54

In Action 60

Approximation and Precedence 65

Significant Figures 68

Quiz 71

CHAPTER 4 **How Things Are Measured** **74**

Systems of Units 75

Base Units in SI 76

Other Units in SI 83

Conversions 87

Quiz 91

Test: Part 1 **93**

CONTENTS

PART 2:	**FINDING UNKNOWNS**	**103**
CHAPTER 5	**Basic Algebra**	**105**
	Single-Variable Linear Equations	105
	Two-by-Two Linear Equations	110
	Quiz	116
CHAPTER 6	**More Algebra**	**119**
	Quadratic Equations	119
	Beyond Reality	129
	One-Variable, Higher-Order Equations	133
	Quiz	140
CHAPTER 7	**A Statistics Sampler**	**143**
	Experiments and Variables	143
	Populations and Samples	146
	Distributions	149
	More Definitions	154
	Quiz	163
CHAPTER 8	**Taking Chances**	**165**
	The Probability Fallacy	165

Definitions   167

Properties of Outcomes   172

Permutations and Combinations   181

Quiz   184

Test: Part 2   **187**

PART 3:   **SHAPES AND PLACES**   **197**

CHAPTER 9   **Geometry on the Flats**   **199**

Fundamental Rules   199

Triangles   208

Quadrilaterals   214

Circles and Ellipses   223

Quiz   226

CHAPTER 10   **Geometry in Space**   **229**

Points, Lines, and Planes   229

Straight-Edged Objects   235

Cones, Cylinders, and Spheres   240

Quiz   247

CONTENTS

CHAPTER 11	**Graphing It**	**250**
	The Cartesian Plane	250
	Straight Lines in the Cartesian Plane	254
	The Polar Coordinate Plane	260
	Some Examples	263
	Quiz	271
CHAPTER 12	**A Taste of Trigonometry**	**274**
	More about Circles	274
	Primary Circular Functions	277
	Secondary Circular Functions	283
	The Right Triangle Model	287
	Pythagorean Extras	290
	Quiz	292
	Test: Part 3	**295**
PART 4:	**MATH IN SCIENCE**	**307**
CHAPTER 13	**Vectors and 3D**	**309**
	Vectors in the Cartesian Plane	309
	Rectangular 3D Coordinates	313

	Vectors in Cartesian Three-Space	316
	Flat Planes in Space	323
	Straight Lines in Space	326
	Quiz	330
CHAPTER 14	**Growth and Decay**	**332**
	Growth by Addition	332
	Growth by Multiplication	338
	Exponential Functions	345
	Rules for Exponentials	347
	Logarithms	350
	Rules for Logarithms	352
	Graphs Based on Logarithms	355
	Quiz	359
CHAPTER 15	**How Things Move**	**362**
	Mass and Force	362
	Displacement	367
	Speed and Velocity	368
	Acceleration	373

CONTENTS

Momentum   378

Quiz   381

Test: Part 4   **384**

Final Exam   **396**

**Answers to Quiz, Test, and
Exam Questions**   **424**

Suggested Additional References   **428**

Index   **431**

PREFACE

This book is for people who want to refine their math skills at the high-school level. It can serve as a supplemental text in a classroom, tutored, or home-schooling environment. It should also be useful for career changers who want to refresh or augment their knowledge. I recommend that you start at the beginning and complete a chapter a week. An hour or two daily ought to be enough time for this. When you're done, you can use this book as a permanent reference.

This course has an abundance of practice quiz, test, and exam questions. They are all multiple-choice, and are similar to the sorts of questions used in standardized tests. There is a short quiz at the end of every chapter. The quizzes are "open-book." You may (and should) refer to the chapter texts when taking them. When you think you're ready, take the quiz, write down your answers, and then give your answers to a friend. Have the friend tell you your score, but not which questions you got wrong. The answers are listed in the back of the book. Stick with a chapter until you get most of the answers correct.

This book is divided into multi-chapter sections. At the end of each section, there is a multiple-choice test. Take these tests when you're done with the respective sections and have taken all the chapter quizzes. The section tests are "closed-book," but the questions are easier than those in the quizzes. A satisfactory score is 75% or more correct. Again, answers are in the back of the book.

There is a final exam at the end of this course. It contains questions drawn uniformly from all the chapters. A satisfactory score is at least 75% correct answers. With the section tests and the final exam, as with the quizzes, have a friend tell you your score without letting you know which questions you missed. That way, you will not subconsciously memorize the answers. You can check to see where your knowledge is strong and where it is not.

When you're finished with this supplemental course, you'll have an advantage over your peers. You'll have the edge when it comes to figuring out solutions to problems in a variety of situations people encounter in today's technological world. You'll understand the "why," as well as the "what" and the "how."

Suggestions for future editions are welcome.

Stan Gibilisco

Acknowledgments

Illustrations in this book were generated with *CorelDRAW*. Some of the clip art is courtesy of Corel Corporation.

I extend thanks to Emma Previato of Boston University, who helped with the technical editing of the manuscript.

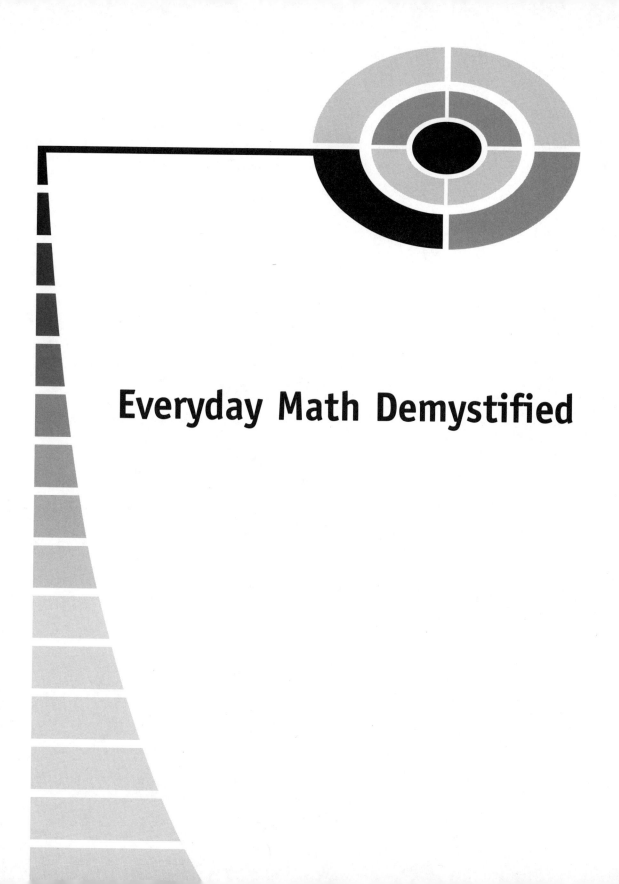

Everyday Math Demystified

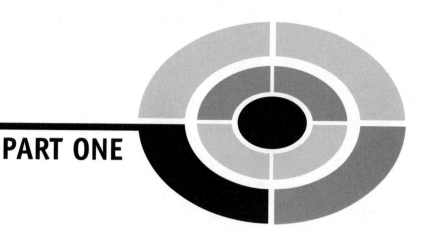

PART ONE

Expressing Quantities

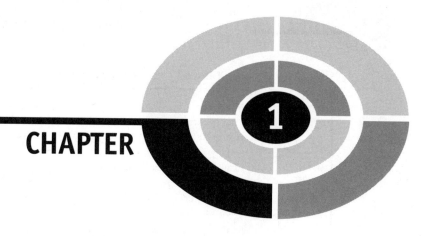

CHAPTER

1

Numbers and Arithmetic

Mathematics is expressed in language alien to people unfamiliar with it. Someone talking about mathematics can sound like a rocket scientist. Written mathematical documents are often laden with symbology. Before you proceed further, look over Table 1-1. It will help you remember symbols used in basic mathematics, and might introduce you to a few symbols you've never seen before!

If at first some of this stuff seems theoretical and far-removed from "the everyday world," think of it as basic training, a sort of math boot camp. Or better yet, think of it as the classroom part of drivers' education. It was good to have that training so you'd know how to read the instrument panel, find the turn signal lever, adjust the mirrors, control the headlights, and read the road signs. So get ready for a drill. Get ready to think logically. Get your mind into math mode.

Table 1-1   Symbols used in basic mathematics.

Symbol	Description
{ }	Braces; objects between them are elements of a set
$\Rightarrow$	Logical implication; read "implies"
$\Longleftrightarrow$	Logical equivalence; read "if and only if"
$\forall$	Universal quantifier; read "for all" or "for every"
$\exists$	Existential quantifier; read "for some"
|	Logical expression; read "such that"
&	Logical conjunction; read "and"
N	The set of natural numbers
Z	The set of integers
Q	The set of rational numbers
R	The set of real numbers
$\varnothing$	The set with no elements; read "the empty set" or "the null set"
$\cap$	Set intersection; read "intersect"
$\cup$	Set union; read "union"
$\subset$	Proper subset; read "is a proper subset of"
$\subseteq$	Subset; read "is a subset of"
$\in$	Element; read "is an element of" or "is a member of"
$\notin$	Non-element; read "is not an element of" or "is not a member of"
$=$	Equality; read "equals" or "is equal to"
$\neq$	Not-equality; read "does not equal" or "is not equal to"

(Continued)

Table 1-1 Continued.

Symbol	Description
≈	Approximate equality; read "is approximately equal to"
<	Inequality; read "is less than"
≤	Weak inequality; read "is less than or equal to"
>	Inequality; read "is greater than"
≥	Weak inequality; read "is greater than or equal to"
+	Addition; read "plus"
−	Subtraction, read "minus"
× · *	Multiplication; read "times" or "multiplied by"
÷ /	Quotient; read "over" or "divided by"
:	Ratio or proportion; read "is to"
	Logical expression; read "such that"
!	Product of all natural numbers from 1 up to a certain value; read "factorial"
( )	Quantification; read "the quantity"
[ ]	Quantification; used outside ( )
{ }	Quantification; used outside [ ]

Sets

A *set* is a collection or group of definable *elements* or *members*. Set elements commonly include:

- points on a line
- instants in time
- coordinates in a plane
- coordinates in space
- coordinates on a display
- curves on a graph or display
- physical objects
- chemical elements
- locations in memory or storage
- data bits, bytes, or characters
- subscribers to a network

If an object or number (call it a) is an element of set A, this fact is written as:

$$a \in A$$

The $\in$ symbol means "is an element of."

SET INTERSECTION

The *intersection* of two sets A and B, written $A \cap B$, is the set C such that the following statement is true for every element x:

$$x \in C \text{ if and only if } x \in A \text{ and } x \in B$$

The $\cap$ symbol is read "intersect."

SET UNION

The *union* of two sets A and B, written $A \cup B$, is the set C such that the following statement is true for every element x:

$$x \in C \text{ if and only if } x \in A \text{ or } x \in B$$

The $\cup$ symbol is read "union."

SUBSETS

A set A is a *subset* of a set B, written $A \subseteq B$, if and only if the following holds true:

$$x \in A \text{ implies that } x \in B$$

The $\subseteq$ symbol is read "is a subset of." In this context, "implies that" is meant in the strongest possible sense. The statement "This implies that" is equivalent to "If this is true, then that is always true."

PROPER SUBSETS

A set A is a *proper subset* of a set B, written $A \subset B$, if and only if the following both hold true:

$$x \in A \text{ implies that } x \in B$$
$$\text{as long as } A \neq B$$

The $\subset$ symbol is read "is a proper subset of."

DISJOINT SETS

Two sets A and B are *disjoint* if and only if all three of the following conditions are met:

$$A \neq \varnothing$$
$$B \neq \varnothing$$
$$A \cap B = \varnothing$$

where $\varnothing$ denotes the *empty set*, also called the *null set*. It is a set that doesn't contain any elements, like a basket of apples without the apples.

COINCIDENT SETS

Two non-empty sets A and B are *coincident* if and only if, for all elements x, both of the following are true:

$$x \in A \text{ implies that } x \in B$$
$$x \in B \text{ implies that } x \in A$$

Numbering Systems

A *number* is an abstract expression of a quantity. Mathematicians define numbers in terms of sets containing sets. All the known numbers can be built up from a starting point of zero. *Numerals* are the written symbols that are agreed-on to represent numbers.

NATURAL AND WHOLE NUMBERS

The *natural numbers*, also called *whole numbers* or *counting numbers*, are built up from a starting point of 0 or 1, depending on which text you consult. The set of natural numbers is denoted N. If we include 0, we have this:

$$N = \{0, 1, 2, 3, \ldots, n, \ldots\}$$

In some instances, 0 is not included, so:

$$N = \{1, 2, 3, 4, \ldots, n, \ldots\}$$

Natural numbers can be expressed as points along a geometric ray or half-line, where quantity is directly proportional to displacement (Fig. 1-1).

DECIMAL NUMBERS

The *decimal number system* is also called *base 10* or *radix 10*. *Digits* in this system are the elements of the set {0, 1, 2, 3, 4, 5, 6, 7, 8, 9}. Numerals are written out as strings of digits to the left and/or right of a *radix point*, which is sometimes called a "decimal point."

In the expression of a decimal number, the digit immediately to the left of the radix point is multiplied by 1, and is called the *ones digit*. The next digit to the left is multiplied by 10, and is called the *tens digit*. To the left of this are digits representing hundreds, thousands, tens of thousands, and so on.

The first digit to the right of the radix point is multiplied by a factor of 1/10, and is called the *tenths digit*. The next digit to the right is multiplied by 1/100, and is called the *hundredths digit*. Then come digits representing thousandths, ten-thousandths, and so on.

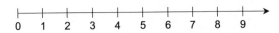

Fig. 1-1. The natural numbers can be depicted as points on a half-line or ray.

BINARY NUMBERS

The *binary number system* is a scheme for expressing numbers using only the digits 0 and 1. It is sometimes called *base 2*. The digit immediately to the left of the radix point is the "ones" digit. The next digit to the left is the "twos" digit; after that comes the "fours" digit. Moving further to the left, the digits represent 8, 16, 32, 64, and so on, doubling every time. To the right of the radix point, the value of each digit is cut in half again and again, that is, 1/2, 1/4, 1/8, 1/16, 1/32, 1/64, and so on.

When you work with a computer or calculator, you give it a decimal number that is converted into binary form. The computer or calculator does its operations with zeros and ones, also called digital low and high states. When the process is complete, the machine converts the result back into decimal form for display.

OCTAL AND HEXADECIMAL NUMBERS

The *octal number system* uses eight symbols. Every digit is an element of the set {0, 1, 2, 3, 4, 5, 6, 7}. Starting with 1, counting in the octal system goes like this: 1, 2, 3, 4, 5, 6, 7, 10, 11, 12, ... 16, 17, 20, 21, ... 76, 77, 100, 101, ... 776, 777, 1000, 1001, and so on.

Yet another scheme, commonly used in computer practice, is the *hexadecimal number system*, so named because it has 16 symbols. These digits are the decimal 0 through 9 plus six more, represented by A through F, the first six letters of the alphabet. Starting with 1, counting in the hexadecimal system goes like this: 0, 1, 2, 3, 4, 5, 6, 7, 8, 9, A, B, C, D, E, F, 10, 11, ... 1E, 1F, 20, 21, ... FE, FF, 100, 101, ... FFE, FFF, 1000, 1001, and so on.

PROBLEM 1-1
What is the value of the binary number 1001011 in the decimal system?

SOLUTION 1-1
Note the decimal values of the digits in each position, proceeding from right to left, and add them all up as a running sum. The right-most digit, 1, is multiplied by 1. The next digit to the left, 1, is multiplied by 2, for a running sum of 3. The digit to the left of that, 0, is multiplied by 4, so the running sum is still 3. The next digit to the left, 1, is multiplied by 8, so the running sum is 11. The next two digits to the left of that, both 0, are multiplied by 16 and 32, respectively, so the running sum is still 11. The left-most digit, 1, is multiplied by 64, for a final sum of 75. Therefore, the decimal equivalent of 1001011 is 75.

PROBLEM 1-2
What is the value of the binary number 1001.011 in the decimal system?

SOLUTION 1-2
To solve this problem, begin at the radix point. Proceeding to the left first, figure out the whole-number part of the expression. The right-most digit, 1, is multiplied by 1. The next two digits to the left, both 0, are multiplied by 2 and 4, respectively, so the running sum is still 1. The left-most digit, 1, is multiplied by 8, so the final sum for the whole-number part of the expression is 9.

Return to the radix point and proceed to the right. The first digit, 0, is multiplied by 1/2, so the running sum is 0. The next digit to the right, 1, is multiplied by 1/4, so the running sum is 1/4. The right-most digit, 1, is multiplied by 1/8, so the final sum for the fractional part of the expression is 3/8.

Adding the fractional value to the whole-number value in decimal form produces the decimal equivalent of 1001.011, which is 9-3/8.

Integers

The set of natural numbers can be duplicated and inverted to form an identical, mirror-image set:

$$-N = \{0, -1, -2, -3, \ldots\}$$

The union of this set with the set of natural numbers produces the set of integers, commonly denoted Z:

$$Z = N \cup -N$$
$$= \{\ldots, -3, -2, -1, 0, 1, 2, 3, \ldots\}$$

POINTS ALONG A LINE

Integers can be expressed as points along a line, where quantity is directly proportional to displacement (Fig. 1-2). In the illustration, integers

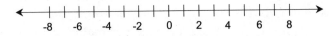

Fig. 1-2. The integers can be depicted as points on a line.

correspond to points where hash marks cross the line. The set of natural numbers is a proper subset of the set of integers:

$$N \subset Z$$

For any number a, if $a \in N$, then $a \in Z$. This is formally written:

$$\forall a: a \in N \Rightarrow a \in Z$$

The converse of this is not true. There are elements of Z (namely, the negative integers) that are not elements of N.

ADDITION OF INTEGERS

Addition is symbolized by the plus sign (+). The result of this operation is a *sum*. Subtraction is symbolized by a long dash (−). The result of this operation is a *difference*. In a sense, these operations "undo" each other. For any integer a, the addition of an integer $-a$ to a quantity is equivalent to the subtraction of the integer a from that quantity. The subtraction of an integer $-a$ from a quantity is equivalent to the addition of the integer a to that quantity.

MULTIPLICATION OF INTEGERS

Multiplication is symbolized by a tilted cross (×), a small dot (·), or sometimes in the case of variables, by listing the numbers one after the other (for example, ab). Occasionally an asterisk (*) is used. The result of this operation is a *product*.

On the number line of Fig. 1-2, products are depicted by moving away from the zero point, or *origin*, either toward the left or toward the right depending on the signs of the numbers involved. To illustrate $a \times b = c$, start at the origin, then move away from the origin a units b times. If a and b are both positive or both negative, move toward the right; if a and b have opposite sign, move toward the left. The finishing point corresponds to c.

The preceding three operations are *closed over the set of integers*. This means that if a and b are integers, then $a + b$, $a - b$, and $a \times b$ are integers. If you add, subtract, and multiply integers by integers, you can never get anything but another integer.

DIVISION OF INTEGERS

Division is symbolized by a forward slash (/) or a dash with dots above and below (÷). The result of this operation is called a *quotient*. When a quotient is expressed as a *ratio* or as a *proportion*, a colon is often used between the numbers involved.

On the number line of Fig. 1-2, quotients are depicted by moving in toward the origin, either toward the left or toward the right depending on the signs of the numbers involved. To illustrate $a/b=c$, it is easiest to envision the product $b \times c = a$ performed "backwards." (Or you can simply use a calculator!)

The operation of division, unlike the operations of addition, subtraction, and multiplication, is not closed over the set of integers. If a and b are integers, then a/b might be an integer, but this is not necessarily the case. Division gives rise to a more comprehensive set of numbers, which we'll look at shortly. The quotient a/b is not defined at all if $b=0$.

EXPONENTIATION WITH INTEGERS

Exponentiation, also called *raising to a power*, is symbolized by a superscript numeral. The result of this operation is known as a *power*.

If a is an integer and b is a positive integer, then a^b is the result of multiplying a by itself b times. For example:

$$2^3 = 2 \times 2 \times 2 = 8$$
$$3^4 = 3 \times 3 \times 3 \times 3 = 81$$

If a is an integer and c is a negative integer, then a^c is the result of multiplying a by itself c times and then dividing 1 by the result. For example,

$$2^{-3} = 1/(2 \times 2 \times 2) = 1/8$$
$$3^{-4} = 1/(3 \times 3 \times 3 \times 3) = 1/81$$

PRIME NUMBERS

Let p be a nonzero natural number. Suppose $ab=p$, where a and b are natural numbers. Further suppose that either of the following statements is true for all a and b:

$$(a = 1) \& (b = p)$$
$$(a = p) \& (b = 1)$$

Then p is defined as a *prime number*. In other words, a natural number p is prime if and only if its only two natural-number factors are 1 and itself.

The first several prime numbers are 1, 2, 3, 5, 7, 11, 13, 17, 19, 23, 29, 31, and 37. Sometimes 1 is not included in this list, although technically it meets the above requirement for "prime-ness."

PRIME FACTORS

Suppose n is a natural number. Then there exists a unique sequence of prime numbers p_1, p_2, p_3,...,p_m, such that both of the following statements are true:

$$p_1 \leq p_2 \leq p_3 \leq \cdots \leq p_m$$
$$p_1 \times p_2 \times p_3 \times \ldots \times p_m = n$$

The numbers $p_1, p_2, p_3, \ldots, p_m$ are called the *prime factors* of the natural number n. Every natural number n has one, but only one, set of prime factors. This is an important principle known as the *Fundamental Theorem of Arithmetic*.

PROBLEM 1-3
What are the prime factors of 80?

SOLUTION 1-3
Here's a useful hint for finding the prime factors of any natural number n: Unless n is a prime number, all of its prime factors are less than or equal to $n/2$. In this case $n = 80$, so we can be certain that all the prime factors of 80 are less than or equal to 40.
   Finding the prime factors is largely a matter of repeating this divide-by-2 process over and over, and making educated guesses to break down the resulting numbers into prime factors. We can see that $80 = 2 \times 40$. Breaking it down further:

$$80 = 2 \times (2 \times 20)$$
$$= 2 \times 2 \times (2 \times 10)$$
$$= 2 \times 2 \times 2 \times 2 \times 5$$

PROBLEM 1-4
Is the number 123 prime? If not, what are its prime factors?

SOLUTION 1-4
The number 123 is not prime. It can be factored into $123 = 41 \times 3$. The numbers 41 and 3 are both prime, so they are the prime factors of 123.

Rational, Irrational, and Real Numbers

The natural numbers, or counting numbers, have plenty of interesting properties. But when we start dividing them by one another or performing fancy operations on them like square roots, trigonometric functions, and logarithms, things can get downright fascinating.

QUOTIENT OF TWO INTEGERS

A *rational number* (the term derives from the word *ratio*) is a quotient of two integers, where the denominator is not zero. The standard form for a rational number a is:

$$a = m/n$$

where m and n are integers, and $n \neq 0$. The set of all possible such quotients encompasses the entire set of rational numbers, denoted $\boldsymbol{Q}$. Thus,

$$\boldsymbol{Q} = \{r \mid r = m/n\}$$

where $m \in \boldsymbol{Z}$, $n \in \boldsymbol{Z}$, and $n \neq 0$. The set of integers is a proper subset of the set of rational numbers. Thus, the natural numbers, the integers, and the rational numbers have the following relationship:

$$N \subset \boldsymbol{Z} \subset \boldsymbol{Q}$$

DECIMAL EXPANSIONS

Rational numbers can be denoted in decimal form as an integer, followed by a radix point, followed by a sequence of digits. (See **Decimal numbers** above for more details concerning this notation.) The digits following the radix point always exist in either of two forms:

- a finite string of digits
- an infinite string of digits that repeat in cycles

Examples of the first type of rational number, known as *terminating decimals*, are:

$$3/4 = 0.750000\ldots$$
$$-9/8 = -1.1250000\ldots$$

Examples of the second type of rational number, known as *nonterminating, repeating decimals*, are:

$$1/3 = 0.33333\ldots$$
$$-1/6 = -0.166666\ldots$$

RATIONAL-NUMBER "DENSITY"

One of the most interesting things about rational numbers is the fact that they are "dense." Suppose we assign rational numbers to points on a line, in such a way that the distance of any point from the origin is directly proportional to its numerical value. If a point is on the left-hand side of the point representing 0, then that point corresponds to a negative number; if it's on the right-hand side, it corresponds to a positive number. If we mark off only the integers on such a line, we get a picture that looks like Fig. 1-2. But in the case of the rational numbers, there are points all along the line, not only at those places where the hash marks cross the line.

If we take any two points *a* and *b* on the line that correspond to rational numbers, then the point midway between them corresponds to the rational number $(a+b)/2$. This is true no matter how many times we repeat the operation. We can keep cutting an interval in half forever, and if the end points are both rational numbers, then the midpoint is another rational number. Figure 1-3 shows an example of this. It is as if you could take a piece

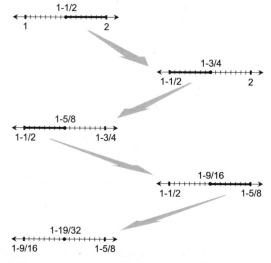

Fig. 1-3.   An interval can be repeatedly cut in half, generating rational numbers without end.

of paper and keep folding it over and over, and never get to the place where you couldn't fold it again.

It is tempting to suppose that points on a line, defined as corresponding to rational numbers, are "infinitely dense." They are, in a sense, squeezed together "infinitely tight" so that every point in between any two points must correspond to some rational number. But do the rational numbers account for all of the points along a true geometric line? The answer, which surprises many folks the first time they hear it, is "No."

IRRATIONAL NUMBERS

An *irrational number* is a number that cannot be expressed as the ratio of two integers. Examples of irrational numbers include:

- the length of the diagonal of a square that is one unit on each edge
- the circumference-to-diameter ratio of a circle.

All irrational numbers share the property of being inexpressible in decimal form. When an attempt is made to express such a number in this form, the result is a *nonterminating, nonrepeating* decimal. No matter how many digits are specified to the right of the radix point, the expression is only an approximation of the actual value of the number. The set of irrational numbers can be denoted S. This set is entirely disjoint from the set of rational numbers. That means that no irrational number is rational, and no rational number is irrational:

$$S \cap Q = \varnothing$$

REAL NUMBERS

The set of *real numbers*, denoted R, is the union of the sets of rational and irrational numbers:

$$R = Q \cup S$$

For practical purposes, R can be depicted as the set of points on a continuous geometric line, as shown in Fig. 1-2. (In theoretical mathematics, the assertion that the points on a geometric line correspond one-to-one with the real numbers is known as the *Continuum Hypothesis*.) The real numbers are related to the rational numbers, the integers, and the natural numbers as follows:

$$N \subset Z \subset Q \subset R$$

The operations of addition, subtraction, multiplication, and division can be defined over the set of real numbers. If # represents any one of these operations and x and y are elements of R with $y \neq 0$, then:

$$x \# y \in R$$

REAL-NUMBER "DENSITY"

Do you sense something strange going on here? We've just seen that the rational numbers, when depicted as points along a line, are "dense." No matter how close together two rational-number points on a line might be, there is always another rational-number point between them. But this doesn't mean that the rational-number points are the only points on a line.

Every rational number corresponds to some point on a number line such as the one shown in Fig. 1-2. But the converse of this statement is not true. There are some points on the line that don't correspond to rational numbers. A good example is the positive number that, when multiplied by itself, produces the number 2. This is approximately equal to 1.41421. Another example is the ratio of the circumference of a circle to its diameter, a constant commonly called *pi* and symbolized π. It's approximately equal to 3.14159.

The set of real numbers is more "dense" than the set of rational numbers. How many times more dense? Twice? A dozen times? A hundred times? It turns out that the set of real numbers, when depicted as the points on a line, is infinitely more dense than the set of real numbers. This is hard to imagine, and a proof of it is beyond the scope of this book. You might think of it this way: Even if you lived forever, you would die before you could name all the real numbers.

SHADES OF INFINITY

The symbol $\aleph_0$ (aleph-null or aleph-nought) denotes the cardinality of the set of rational numbers. The cardinality of the real numbers is denoted $\aleph_1$ (aleph-one). These "numbers" are called *infinite cardinals* or *transfinite cardinals*.

Around the year 1900, the German mathematician Georg Cantor proved that these two "numbers" are not the same. The infinity of the real numbers is somehow larger than the infinity of the rational numbers:

$$\aleph_1 > \aleph_0$$

The elements of N can be paired off one-to-one with the elements of Z or Q, but not with the elements of S or R. Any attempt to pair off the elements of N

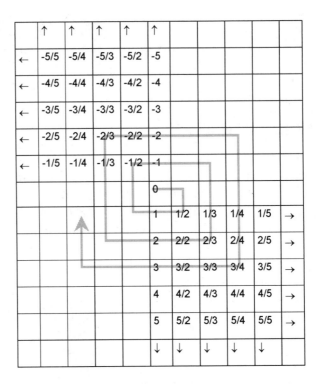

	↑	↑	↑	↑	↑					
←	-5/5	-5/4	-5/3	-5/2	-5					
←	-4/5	-4/4	-4/3	-4/2	-4					
←	-3/5	-3/4	-3/3	-3/2	-3					
←	-2/5	-2/4	-2/3	-2/2	-2					
←	-1/5	-1/4	-1/3	-1/2	-1					
					0					
					1	1/2	1/3	1/4	1/5	→
					2	2/2	2/3	2/4	2/5	→
					3	3/2	3/3	3/4	3/5	→
					4	4/2	4/3	4/4	4/5	→
					5	5/2	5/3	5/4	5/5	→
					↓	↓	↓	↓	↓	

Fig. 1-4. A tabular listing scheme for the rational numbers. Proceed as shown by the gray line. If a box is empty, or if the number in it is equivalent to one encountered previously (such as 3/3 or −2/4), skip the box without counting it.

and S or N and R results in some elements of S or R being left over without corresponding elements in N. This reflects the fact that the elements of N, Z, or Q can be defined in terms of a listing scheme (Fig. 1-4 is an example), but the elements of S or R cannot. It also reflects the fact that the points on a real-number line are more "dense" than the points on a line denoting the natural numbers, the integers, or the rational numbers.

PROBLEM 1-5
What is 4/7, expressed using only a radix point and decimal digits?

SOLUTION 1-5
You can divide this out "longhand," or you can use a calculator that has a display with a lot of digits. The result is:

$$4/7 = 0.571428571428571428\ldots$$

That is zero, followed by a radix point, followed by the sequence of digits 571428 repeating over and over without end.

PROBLEM 1-6
What is the value of 3.367367367... expressed as a fraction?

SOLUTION 1-6
First consider only the portion to the right of the radix point. This is a repeating sequence of the digits 367, over and over without end. Whenever you see a repeating sequence of several digits to the right of a radix point, its fractional equivalent is found by dividing that sequence of digits by an equal number of nines. In this case:

$$0.367367367... = 367/999$$

To get the value of 3.367367367... as a fraction, simply add 3 to the above, getting:

$$3.367367367... = 3\text{-}367/999$$

In this context, the dash between the whole number portion and the fractional portion of the expression serves only to separate them for notational clarity. (It isn't a minus sign!)

Number Operations

Several properties, also called principles or laws, are recognized as valid for the operations of addition, subtraction, multiplication, and division for all real numbers. Here are some of them. It's not a bad idea to memorize these. You probably learned them in elementary school.

ADDITIVE IDENTITY ELEMENT

When 0 is added to any real number a, the sum is always equal to a. The number 0 is said to be the *additive identity element*:

$$a + 0 = a$$

MULTIPLICATIVE IDENTITY ELEMENT

When any real number a is multiplied by 1, the product is always equal to a. The number 1 is said to be the *multiplicative identity element*:

$$a \times 1 = a$$

ADDITIVE INVERSES

For every real number a, there exists a unique real number $-a$ such that the sum of the two is equal to 0. The numbers a and $-a$ are called *additive inverses*:

$$a + (-a) = 0$$

MULTIPLICATIVE INVERSES

For every nonzero real number a, there exists a unique real number $1/a$ such that the product of the two is equal to 1. The numbers a and $1/a$ are called *multiplicative inverses*:

$$a \times (1/a) = 1$$

The multiplicative inverse of a real number is also called its *reciprocal*.

COMMUTATIVE LAW FOR ADDITION

When any two real numbers are added together, it does not matter in which order the sum is performed. The operation of addition is said to be *commutative over the set of real numbers*. For all real numbers a and b, the following equation is valid:

$$a + b = b + a$$

COMMUTATIVE LAW FOR MULTIPLICATION

When any two real numbers are multiplied by each other, it does not matter in which order the product is performed. The operation of multiplication, like addition, is commutative over the set of real numbers. For all real numbers a and b, the following equation is always true:

$$a \times b = b \times a$$

A product can be written without the "times sign" ($\times$) if, but only if, doing so does not result in an ambiguous or false statement. The above expression is often seen written this way:

$$ab = ba$$

If $a = 3$ and $b = 52$, however, it's necessary to use the "times sign" and write this:

$$3 \times 52 = 52 \times 3$$

The reason becomes obvious if the above expression is written without using the "times signs." This results in a false statement:

$$352 = 523$$

ASSOCIATIVE LAW FOR ADDITION

When adding any three real numbers, it does not matter how the addends are grouped. The operation of addition is *associative over the set of real numbers*. For all real numbers a_1, a_2, and a_3, the following equation holds true:

$$(a_1 + a_2) + a_3 = a_1 + (a_2 + a_3)$$

ASSOCIATIVE LAW FOR MULTIPLICATION

When multiplying any three real numbers, it does not matter how the multiplicands are grouped. Multiplication, like addition, is associative over the set of real numbers. For all real numbers a, b, and c, the following equation holds:

$$(ab)c = a(bc)$$

DISTRIBUTIVE LAWS

For all real numbers a, b, and c, the following equation holds. The operation of multiplication is *distributive with respect to addition*:

$$a(b + c) = ab + ac$$

The above statement logically implies that multiplication is distributive with respect to subtraction, as well:

$$a(b - c) = ab - ac$$

The distributive law can also be extended to division as long as there aren't any denominators that end up being equal to zero. For all real numbers a, b, and c, where $a \neq 0$, the following equations are valid:

$$(ab + ac)/a = ab/a + ac/a = b + c$$
$$(ab - ac)/a = ab/a - ac/a = b - c$$

PRECEDENCE OF OPERATIONS

When addition, subtraction, multiplication, division, and exponentiation (raising to a power) appear in an expression and that expression must be simplified, the operations should be performed in the following sequence:

- Simplify all expressions within parentheses, brackets, and braces from the inside out.
- Perform all exponential operations, proceeding from left to right.
- Perform all products and quotients, proceeding from left to right.
- Perform all sums and differences, proceeding from left to right.

The following are examples of this process, in which the order of the numerals and operations is the same in each case, but the groupings differ:

$$[(2+3)(-3-1)^2]^2$$
$$= [5 \times (-4)^2]^2$$
$$= (5 \times 16)^2$$
$$= 80^2 = 6400$$

$$[(2+3 \times (-3)-1)^2]^2$$
$$= [(2+(-9)-1)^2]^2$$
$$= (-8^2)^2$$
$$= 64^2 = 4096$$

A note of caution is in order here: This rule doesn't apply to exponents of exponents. For example, 3^3 raised to the power of 3 is equal to 27^3 or 19,683. But 3 raised to the power of 3^3 is equal to 3^{27} or 7,625,597,484,987.

PROBLEM 1-7
What is the value of the expression $-5 \times 6 + 8 - 1$?

SOLUTION 1-7
First, multiply -5 times 6, getting the product -30. Then add 8 to this, getting -22. Finally subtract 1, getting -23. Thus:

$$-5 \times 6 + 8 - 1 = -23$$

PROBLEM 1-8

What is the value of the expression $-5 \times (6 + 8 - 1)$?

SOLUTION 1-8

First find the sum $6 + 8$, which is 14. Then subtract 1 from this, getting 13. Finally multiply 13 by -5, getting -65. Thus:

$$-5 \times (6 + 8 - 1) = -65$$

More Principles Worth Memorizing

The following rules and definitions apply to arithmetic operations for real numbers, with the constraint that no denominator can ever be equal to zero. It's good to memorize these, because an "automatic" knowledge of them can save you some time as you hack your way through the occasional forest of calculations that life is bound to confront you with. After a while, these laws will seem like nothing more or less than common sense.

ZERO NUMERATOR

For all nonzero real numbers a, if 0 is divided by a, then the result is equal to 0. This equation is always true:

$$0/a = 0$$

ZERO DENOMINATOR

For all real numbers a, if a is divided by 0, then the result $a/0$ is undefined. Perhaps a better way of saying this is that, whatever such an "animal" might happen to be, it is not a real number! Within the set of real numbers, there's no such thing as the ratio of a number or variable to 0.

If you're into playing around with numbers, you can have lots of fun trying to define "division by 0" and explore the behavior of ratios with zero denominators. If you plan to go on such a safari, however, take note: The "animals" can be unpredictable, and you had better fortify your mind with lots of ammunition and heavy armor.

MULTIPLICATION BY ZERO

Whenever a real number is multiplied by 0, the product is equal to 0. The following equations are true for all real numbers a:

$$a \times 0 = 0$$
$$0 \times a = 0$$

ZEROTH POWER

Whenever a nonzero real number is taken to the zeroth power, the result is equal to 1. The following equation is true for all real numbers a except 0:

$$a^0 = 1$$

ARITHMETIC MEAN

Suppose that $a_1, a_2, a_3, \ldots, a_n$ be real numbers. The *arithmetic mean*, denoted m_A (also known as the *average*) of $a_1, a_2, a_3, \ldots, a_n$ is equal to the sum of the numbers, divided by the number of numbers. Mathematically:

$$m_A = (a_1 + a_2 + a_3 + \ldots + a_n)/n$$

PRODUCT OF SIGNS

When two real numbers with plus signs (meaning positive, or larger than 0) or minus signs (meaning negative, or less than 0) are multiplied by each other, the following rules apply:

$$(+)(+) = (+)$$
$$(+)(-) = (-)$$
$$(-)(+) = (-)$$
$$(-)(-) = (+)$$

QUOTIENT OF SIGNS

When two real numbers with plus or minus signs are divided by each other, the following rules apply:

$$(+)/(+) = (+)$$
$$(+)/(-) = (-)$$
$$(-)/(+) = (-)$$
$$(-)/(-) = (+)$$

POWER OF SIGNS

When a real number with a plus sign or a minus sign is raised to a positive integer power n, the following rules apply:

$$(+)^n = (+)$$
$$(-)^n = (-) \text{ if } n \text{ is odd}$$
$$(-)^n = (+) \text{ if } n \text{ is even}$$

RECIPROCAL OF RECIPROCAL

For all nonzero real numbers, the reciprocal of the reciprocal is equal to the original number. The following equation holds for all real numbers a provided that $a \neq 0$:

$$1/(1/a) = a$$

PRODUCT OF SUMS

For all real numbers a, b, c, and d, the product of $(a+b)$ with $(c+d)$ is given by the following formula:

$$(a+b)(c+d) = ac + ad + bc + bd$$

CROSS MULTIPLICATION

Given two quotients or ratios expressed as fractions, the numerator of the first times the denominator of the second is equal to the denominator of the first times the numerator of the second. Mathematically, for all real numbers a, b, c, and d where neither a nor b is equal to 0, the following statement is valid:

$$a/b = c/d \Leftrightarrow ad = bc$$

RECIPROCAL OF PRODUCT

For any two nonzero real numbers, the reciprocal (or minus-one power) of their product is equal to the product of their reciprocals. If a and b are both nonzero real numbers:

$$1/(ab) = (1/a)(1/b)$$
$$(ab)^{-1} = a^{-1}b^{-1}$$

PRODUCT OF QUOTIENTS

The product of two quotients or ratios, expressed as fractions, is equal to the product of their numerators divided by the product of their denominators. For all real numbers a, b, c, and d, where $b \neq 0$ and $d \neq 0$:

$$(a/b)(c/d) = (ac)/(bd)$$

RECIPROCAL OF QUOTIENT

For any two nonzero real numbers, the reciprocal (or minus-one power) of their quotient or ratio is equal to the quotient inverted or the ratio reversed. If a and b are real numbers where $a \neq 0$ and $b \neq 0$:

$$1/(a/b) = b/a$$
$$(a/b)^{-1} = b/a$$

Still More Principles

The following general laws of arithmetic can be useful from time to time. You don't really have to memorize them, as long as you can look them up when you need them. (But if you want to memorize them, go right ahead.)

QUOTIENT OF PRODUCTS

The quotient or ratio of two products can be rewritten as a product of two quotients or ratios. For all real numbers a, b, c, and d, where $cd \neq 0$:

$$(ab)/(cd) = (a/c)(b/d)$$

$$= (a/d)(b/c)$$

QUOTIENT OF QUOTIENTS

The quotient or ratio of two quotients or ratios can be rewritten either as a product of two quotients or ratios, or as the quotient or ratio of two products. For all real numbers a, b, c, and d, the following equations hold as long as $b \neq 0$, $c \neq 0$, and $d \neq 0$:

$$(a/b)/(c/d) = (a/b)(d/c)$$

$$= (a/c)(d/b)$$

$$= (ad)/(bc)$$

SUM OF QUOTIENTS (COMMON DENOMINATOR)

If two quotients or ratios have the same nonzero denominator, their sum is equal to the sum of the numerators, divided by the denominator. For all real numbers a, b, and c, where $c \neq 0$:

$$a/c + b/c = (a+b)/c$$

SUM OF QUOTIENTS (GENERAL)

If two quotients or ratios have different nonzero denominators, their sum must be found using a specific formula. For all real numbers a, b, c, and d, where $b \neq 0$ and $d \neq 0$:

$$a/b + c/d = (ad + bc)/(bd)$$

INTEGER ROOTS

Suppose that a is a positive real number. Also suppose that n is a positive integer. Then the nth root of a can also be expressed as the $1/n$ power of a. Thus, the second root (or square root) is the same thing as the 1/2 power; the third root (or cube root) is the same thing as the 1/3 power; the fourth root is the same thing as the 1/4 power; and so on.

RATIONAL-NUMBER POWERS

Suppose that a is a real number. Also suppose that b is a rational number such that $b = m/n$, where m and n are integers and $n \neq 0$. Then the following formula holds:

$$a^b = a^{m/n} = (a^m)^{1/n} = (a^{1/n})^m$$

NEGATIVE POWERS

Suppose that a is a real number. Also suppose that b is a rational number. Then the following formula holds:

$$a^{-b} = (1/a)^b = 1/(a^b)$$

SUM OF POWERS

Suppose that a is a real number. Also suppose that b and c are rational numbers. Then the following formula holds:

$$a^{(b+c)} = a^b a^c$$

DIFFERENCE OF POWERS

Suppose a is a nonzero real number. Also suppose that b and c are rational numbers. Then the following formula holds:

$$a^{(b-c)} = a^b / a^c$$

LET $a = x$, $b = y$, AND $c = z$

In mathematical literature, the expression "Let such-and-such be the case" is often used instead of "Suppose that such-and-such is true" or "Imagine such-and-such." When you are admonished to "let" things be a certain way, you are in effect being asked to imagine, or suppose, that they are so. That's the way the next few principles are stated. Also, different variables are used in the next few statements: x replaces a, y replaces b, and z replaces c. It's a good idea to get used to this sort of variability (along with repetitiveness, ironically) to arm your mind against the assaults of mathematical discourse.

PRODUCT OF POWERS

Let x be a real number. Let y and z be rational numbers. Then the following formula holds:

$$x^{yz} = (x^y)^z = (x^z)^y$$

QUOTIENT OF POWERS

Let x be a real number. Let y and z be rational numbers, with the constraint that $z \neq 0$. Then the following formula holds:

$$x^{y/z} = (x^y)^{1/z}$$

POWER OF RECIPROCAL

Let x be a nonzero real number. Let n be a natural number. Then the nth power of the reciprocal of x is equal to the reciprocal of x to the nth power. That is:

$$(1/x)^n = 1/(x^n)$$

POWERS OF SUM

Let x and y be real numbers. Then the following formulas hold:

$$(x+y)^2 = x^2 + 2xy + y^2$$
$$(x+y)^3 = x^3 + 3x^2y + 3xy^2 + y^3$$
$$(x+y)^4 = x^4 + 4x^3y + 6x^2y^2 + 4xy^3 + y^4$$

POWERS OF DIFFERENCE

Let x and y be real numbers. Then the following formulas hold:

$$(x-y)^2 = x^2 - 2xy + y^2$$
$$(x-y)^3 = x^3 - 3x^2y + 3xy^2 - y^3$$
$$(x-y)^4 = x^4 - 4x^3y + 6x^2y^2 - 4xy^3 + y^4$$

PROBLEM 1-9
What is the value of $(1/27)^{1/3}$?

SOLUTION 1-9
We are seeking the 1/3 power (or the cube root) of 1/27. This means we are looking for some number which, when multiplied by itself and then multiplied by itself again, will produce 1/27.

Imagine some number y such that $(1/y)^3 = (1/y)(1/y)(1/y) = 1/27$. Then y must be the cube root of 27. Using a calculator, the cube root of 27 is found to be 3. From this we can conclude that $1/y = 1/3$, and thus that $(1/3)^3 = 1/27$. The value we seek, the cube root (or 1/3 power) of 1/27, is therefore 1/3.

PROBLEM 1-10
Find the sum of 2/7 and $-5/8$.

SOLUTION 1-10
Use the formula for the sum of quotients in general, stated earlier in this chapter. As long as b and d are nonzero, the following holds for all real numbers a, b, c, and d:

$$a/b + c/d = (ad + bc)/(bd)$$

Assign $a = 2, b = 7, c = -5$, and $d = 8$. Then plug the numbers into the formula:

$$2/7 + (-5/8) = [(2 \times 8) + (7 \times -5)]/(7 \times 8)$$
$$= [16 + (-35)]/56$$
$$= -19/56$$

Quiz

Refer to the text in this chapter if necessary. A good score is eight correct. Answers are in the back of the book.

1. The reciprocal of a number is the same thing as the
    (a) additive inverse
    (b) exponent
    (c) minus-one power
    (d) cube root

2. What is the value of $-2^4 \times 3 + 2$?
    (a) -46
    (b) 50
    (c) 80
    (d) This expression is ambiguous without parentheses to tell us the precedence of the operations.

3. The sum of $3/x$ and $5/x$, where x represents any real number other than 0, is
    (a) $8/(x^2)$
    (b) $8/(2x)$
    (c) $8/x$
    (d) $15/x$

4. The square of $b/4$, where b represents any rational number, is
   (a) $b^2/4$
   (b) $b^2/8$
   (c) $b^2/16$
   (d) $b^b/256$

5. Figure 1-5 is called a *Venn diagram*. It denotes sets as circular regions (including their interiors) on a surface. What does the shaded region in this illustration represent?
   (a) $A+B$
   (b) $A \cap B$
   (c) $A \cup B$
   (d) $A \subseteq B$

6. In Fig. 1-5, the objects labeled P, Q, and R represent individual points. Which of the following statements is true?
   (a) $P \in (A \cup B)$
   (b) $Q \in (A \cap B)$
   (c) $R \in \emptyset$
   (d) $P \subset A$

7. Suppose q, r, and s represent real numbers. Which of the following statements is not true in general, unless constraints are imposed on one or more of the variables?
   (a) $q^{(r+s)} = q^r q^s$
   (b) $q/s - r/s = (q-r)/s$
   (c) $(q+r)+s = q+(r+s)$
   (d) $(qr)s = q(rs)$

8. When we write 34.34343434..., we are writing a representation of
   (a) an irrational number
   (b) a nonterminating, nonrepeating decimal
   (c) a nondenumerable number
   (d) a rational number

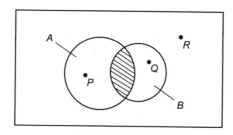

Fig. 1-5. Illustration for Quiz Questions 5 and 6.

9. The 1/2 power of a number is the same thing as
    (a) the square root of that number
    (b) half of that number
    (c) the number minus 1/2
    (d) nothing; it is not defined

10. Suppose x, y, and z are positive real numbers. What is the value of $x^{(y-z)}$ when $y = z$?
    (a) 0
    (b) 1
    (c) x
    (d) It is not defined.

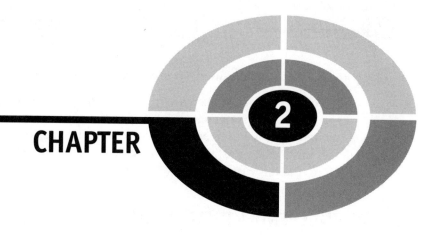

CHAPTER

2

How Variables Relate

Tables and graphs portray relationships between changeable quantities known as *variables*. Tables and graphs show how the values of variables compare with, and in some cases affect, one another. When a table or graph is well composed, it can reveal phenomena that would otherwise be impossible to see.

This versus That

Consider the following statements. Each of them represents a situation that could occur in everyday life.

- The outdoor air temperature varies with the time of day.
- The time the sun is above the horizon on June 21 varies with the latitude of the observer.

● The time required for a wet rag to dry depends on the air temperature.

All of these expressions involve something that depends on something else. In the first case, a statement is made concerning temperature versus time; in the second case, a statement is made concerning sun-up time versus latitude; in the third case, a statement is made concerning time versus temperature. Here, the term *versus* means "depending on."

INDEPENDENT VARIABLES

An *independent variable* can change in value, but its value is not influenced by anything else in a given scenario. Time is often treated as an independent variable. A lot of things depend on time.

When two or more variables are interrelated, at least one of the variables is independent, but they are not all independent. A common and simple situation is one in which there are two variables, one of which is independent. In the three situations described above, the independent variables are time, latitude, and air temperature.

DEPENDENT VARIABLES

A *dependent variable* can change in value, but its value is affected by at least one other factor in a situation. In the scenarios described above, the air temperature, the sun-up time, and time are dependent variables.

When two or more variables are interrelated, at least one of them is dependent, but they cannot all be dependent. Something that's an independent variable in one instance can be a dependent variable in another case. For example, the air temperature is a dependent variable in the first situation described above, but it is an independent variable in the third situation.

SCENARIOS ILLUSTRATED

The three scenarios described above lend themselves to illustration. In order, they are shown crudely in Fig. 2-1.

Figure 2-1A shows an example of outdoor air temperature versus time of day. Drawing B shows the sun-up time (the number of hours per day in which the sun is above the horizon) versus latitude on June 21, where points south of the equator have negative latitude and points north of the equator have positive latitude. Drawing C shows the time it takes for a rag to dry, plotted against the air temperature.

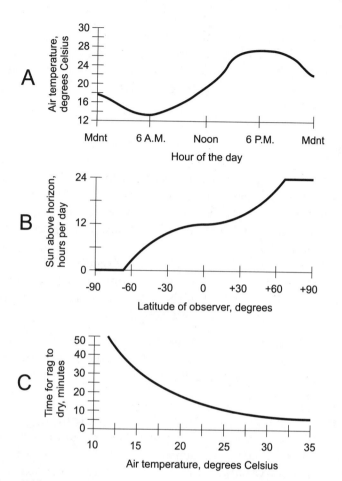

Fig. 2-1.   Three "this-versus-that" scenarios. At A, air temperature versus time of day; at B, sun-up time versus latitude; at C, time for rag to dry versus air temperature.

The scenarios represented by Figs. 2-1A and C are fiction, having been contrived for this discussion. But Fig. 2-1B represents a physical reality; it is true astronomical data for June 21 of every year on earth.

RELATIONS

All three of the graphs in Fig. 2-1 represent *relations*. In mathematics, a relation is an expression of the way two or more variables compare or interact. (It could just as well be called a relationship, a comparison, or an interaction.) Figure 2-1B, for example, is a graph of the relation between the latitude and the sun-up time on June 21.

When dealing with relations, the statements are equally valid if the variables are stated the other way around. Thus, Fig. 2-1B shows a relation between the sun-up time on June 21 and the latitude. In a relation, "this versus that" means the same thing as "that versus this." Relations can always be expressed in graphical form.

FUNCTIONS

A *function* is a special type of mathematical relation. A relation describes how variables compare with each other. In a sense, it is "passive." A function transforms, processes, or morphs the quantity represented by the independent variable into the quantity represented by the dependent variable. A function is "active."

All three of the graphs in Fig. 2-1 represent functions. The changes in the value of the independent variable can, in some sense, be thought of as causative factors in the variations of the value of the dependent variable. We might re-state the scenarios this way to emphasize that they are functions:

- The outdoor air temperature is a function of the time of day.
- The sun-up time on June 21 is a function of the latitude of the observer.
- The time required for a wet rag to dry is a function of the air temperature.

A relation can be a function only when every element in the set of its independent variables has at most one correspondent in the set of dependent variables. If a given value of the dependent variable in a relation has more than one independent-variable value corresponding to it, then that relation might nevertheless be a function. But if any given value of the independent variable corresponds to more than one dependent-variable value, that relation is not a function.

REVERSING THE VARIABLES

In graphs of functions, independent variables are usually represented by horizontal axes, and dependent variables are usually represented by vertical axes. Imagine a movable, vertical line in a graph, and suppose that you can move it back and forth. A curve represents a function if and only if it never intersects the movable vertical line at more than one point.

Imagine that the independent and dependent variables of the functions shown in Fig. 2-1 are reversed. This results in some weird assertions:

- The time of day is a function of the outdoor air temperature.

- The latitude of an observer is a function of the sun-up time on June 21.
- The air temperature is a function of the time it takes for a wet rag to dry.

The first two of these statements are clearly ridiculous. Time does not depend on temperature. You can't make time go backwards by cooling things off or make it rush into the future by heating things up. Your geographic location is not dependent on how long the sun is up. If that were true, you would be at a different latitude a week from now than you are today, even if you don't go anywhere (unless you live on the equator!).

If you turn the graphs of Figs. 2-1A and B sideways to reflect the transposition of the variables and then perform the vertical-line test, you'll see that they no longer depict functions. So the first two of the above assertions are not only absurd; they are false.

Figure 2-1C represents a function, at least in theory, when "stood on its ear." The statement is still strange, but it can at least be true under certain conditions. The drying time of a standard-size wet rag made of a standard material could be used to infer air temperature experimentally (although humidity and wind speed would be factors too). When you want to determine whether or not a certain graph represents a mathematical function, use the vertical-line test, not the common-sense test!

PROBLEM 2-1

Figure 2-2 represents fluctuations in the prices of two hypothetical stocks, called Stock X and Stock Y, over a portion of a business day. Do either or both of these curves represent stock price as functions of time?

SOLUTION 2-1

Both of the curves represent stock price as functions of time. You can determine this using the vertical-line test. Neither of the curves intersects a movable, vertical line more than once. Thus, both curves represent functions of the independent variable (time).

PROBLEM 2-2

Suppose, in the situation shown by Fig. 2-2, the stock price is considered the independent variable, and time is considered the dependent variable. Do either or both of the curves represent time as functions of stock price?

SOLUTION 2-2

To determine this, plot the graphs by "standing the curves on their ears," as shown in Fig. 2-3. (The curves are rotated 90 degrees counterclockwise,

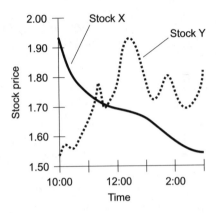

Fig. 2-2. Illustration for Problems 2-1 and 2-2.

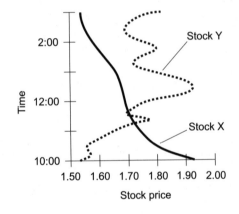

Fig. 2-3. Another illustration for Problem 2-2.

and then mirrored horizontally.) Using the vertical-line test, it is apparent that time can be considered a function of the price of Stock X, but not a function of the price of Stock Y.

Simple Graphs

When the variables in a function can attain only a few values (called *discrete values*), graphs can be rendered in simplified forms. Here are some of the most common types.

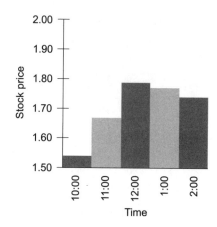

Fig. 2-4.   Vertical bar graph of hypothetical stock price versus time.

VERTICAL BAR GRAPHS

In a *vertical bar graph*, the independent variable is shown on the horizontal axis and the dependent variable is shown on the vertical axis. Function values are portrayed as the heights of bars having equal widths. Figure 2-4 is a vertical bar graph of the price of the hypothetical Stock Y at intervals of 1 hour.

HORIZONTAL BAR GRAPHS

In a *horizontal bar graph*, the independent variable is shown on the vertical axis and the dependent variable is shown on the horizontal axis. Function values are portrayed as the widths of bars having equal heights. Figure 2-5 is a horizontal bar graph of the price of the hypothetical Stock Y at intervals of 1 hour. This type of graph is used much less often than the vertical bar graph. This is because it is customary to place the independent variable on the horizontal axis and the dependent variable on the vertical axis in any graph, when possible.

HISTOGRAMS

A *histogram* is a bar graph applied to a special situation called a *distribution*. An example is a portrayal of the grades a class receives on a test, such as is shown in Fig. 2-6. Here, each vertical bar represents a letter grade (A, B, C, D, or F). The height of the bar represents the percentage of students in the class receiving that grade.

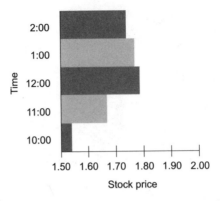

Fig. 2-5. Horizontal bar graph of hypothetical stock price versus time.

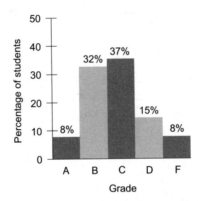

Fig. 2-6. A histogram is a bar graph that shows a statistical distribution.

In Fig. 2-6, the values of the dependent variable are written at the top of each bar. In this case, the percentages add up to 100%, based on the assumption that all of the people in the class are present, take the test, and turn in their papers. The values of the dependent variable are annotated this way in some bar graphs. It's a good idea to write in these numbers if there aren't too many bars in the graph, but it can make the graph look messy or confusing if there are a lot of bars.

In some bar graphs showing percentages, the values do not add up to 100%. We'll see an example of this shortly.

NOMOGRAPHS

A *nomograph* is a one-dimensional graph that consists of two graduated scales lined up directly with each other. Such graphs are useful when the

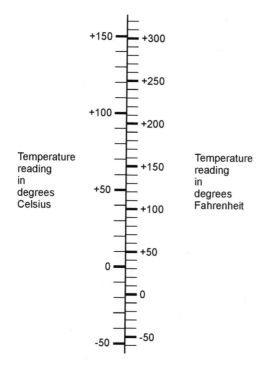

Fig. 2-7.   An example of a nomograph for converting temperature readings.

magnitude of a specific quantity must be compared according to two different unit scales. In Fig. 2-7, a nomograph compares temperature readings in degrees Celsius (also called centigrade) and degrees Fahrenheit.

POINT-TO-POINT GRAPHS

In a *point-to-point graph*, the scales are similar to those used in continuous-curve graphs such as Figs. 2-2 and 2-3. But the values of the function in a point-to-point graph are shown only for a few selected points, which are connected by straight lines.

In the point-to-point graph of Fig. 2-8, the price of Stock Y (from Fig. 2-2) is plotted on the half-hour from 10:00 A.M. to 3:00 P.M. The resulting "curve" does not exactly show the stock prices at the in-between times. But overall, the graph is a fair representation of the fluctuation of the stock over time.

When plotting a point-to-point graph, a certain minimum number of points must be plotted, and they must all be sufficiently close together. If a

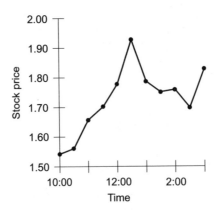

Fig. 2-8.   A point-to-point graph of hypothetical stock price versus time.

point-to-point graph showed the price of Stock Y at hourly intervals, it would not come as close as Fig. 2-8 to representing the actual moment-to-moment stock-price function. If a point-to-point graph showed the price at 15-minute intervals, it would come closer than Fig. 2-8 to the moment-to-moment stock-price function.

CHOOSING SCALES

When composing a graph, it's important to choose sensible scales for the dependent and independent variables. If either scale spans a range of values much greater than necessary, the *resolution* (detail) of the graph will be poor. If either scale does not have a large enough span, there won't be enough room to show the entire function; some of the values will be "cut off."

PROBLEM 2-3
Figure 2-9 is a hypothetical bar graph showing the percentage of the work force in a certain city that calls in sick on each day during a particular work week. What, if anything, is wrong with this graph?

SOLUTION 2-3
The dependent-variable scale is too large. It would be better if the horizontal scale showed values only in the range of 0 to 10%. The graph could also be improved by listing percentage numbers at the right-hand side of each bar. Another way to improve the graph would be to put the independent variable on the horizontal axis and the dependent variable on the vertical axis, making the graph a vertical bar graph instead of a horizontal bar graph.

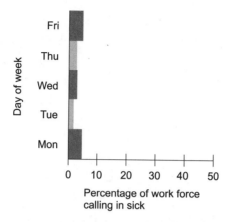

Fig. 2-9.   Illustration for Problems 2-3 and 2-4.

PROBLEM 2-4

What's going on with the percentage values depicted in Fig. 2-9? It is apparent that the values don't add up to 100%. Shouldn't they?

SOLUTION 2-4

No. If they did, it would be a coincidence (and a bad reflection on the attitude of the work force in that city during that week). This is a situation in

n a bar graph does not have to be
ork every day for the whole week, the
d Fig. 2-9 would be perfectly legitimate

s, Trends, and Correlation

fied by "tweaking." Certain charac-
nds and correlation. Here are a few

ween." When a graph is incomplete,
p(s) in order to make the graph
n Fig. 2-10. This is a graph of the
Fig. 2-2, but there's a gap during

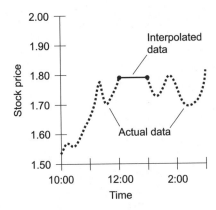

Fig. 2-10.   An example of linear interpolation. The thin solid line represents the interpolation of the values for the gap in the actual available data (heavy dashed curve).

the noon hour. We don't know exactly what happened to the stock price during that hour, but we can fill in the graph using *linear interpolation*. A straight line is placed between the end points of the gap, and then the graph looks complete.

Linear interpolation almost always produces a somewhat inaccurate result. But sometimes it is better to have an approximation than to have no data at all. Compare Fig. 2-10 with Fig. 2-2, and you can see that the *linear interpolation error* is considerable in this case.

CURVE FITTING

Curve fitting is an intuitive scheme for approximating a point-to-point graph, or filling in a graph containing one or more gaps, to make it look like a continuous curve. Figure 2-11 is an approximate graph of the price of hypothetical Stock Y, based on points determined at intervals of half an hour, as generated by curve fitting. This does not precisely represent the actual curve of Fig. 2-2, but it comes close most of the time.

Curve fitting becomes increasingly accurate as the values are determined at more and more frequent intervals. When the values are determined infrequently, this scheme can be subject to large errors, as is shown by the example of Fig. 2-12.

EXTRAPOLATION

The term *extrapolate* means "to put outside of." When a function has a continuous-curve graph where time is the independent variable, *extrapolation*

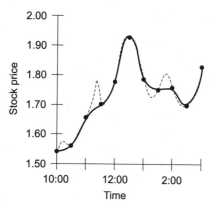

Fig. 2-11.   Approximation of hypothetical stock price as a continuous function of time, making use of curve fitting. The solid curve represents the approximation; the dashed curve represents the actual price as a function of time.

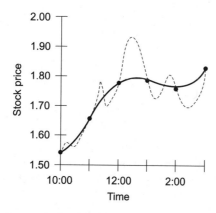

Fig. 2-12.   An example of curve fitting in which not enough data samples are taken, causing significant errors. The solid line represents the approximation; the dashed curve represents the actual price as a function of time.

is the same thing as short-term forecasting. Two examples are shown in Fig. 2-13.

In Fig. 2-13A, the price of the hypothetical Stock X from Fig. 2-2 is plotted until 2:00 P.M., and then an attempt is made to forecast its price for an hour into the future, based on its past performance. In this case, *linear extrapolation*, the simplest form, is used. The curve is simply projected ahead as a straight line. Compare this graph with Fig. 2-2. In this case, linear extrapolation works fairly well.

Figure 2-13B shows the price of the hypothetical Stock Y (from Fig. 2-2) plotted until 2:00 P.M., and then linear extrapolation is used in an attempt to

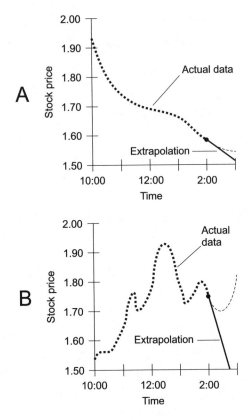

Fig. 2-13. Examples of linear extrapolation. The solid lines represent the forecasts; the dashed curves represent the actual data. In the case shown at A, the prediction is fairly good. In the case shown at B, the linear extrapolation is way off.

predict its behavior for the next hour. As you can see by comparing this graph with Fig. 2-2, linear extrapolation does not work well in this scenario.

Extrapolation is best done by computers. Machines can notice subtle characteristics of functions that humans miss. Some graphs are easy to extrapolate, and others are not. In general, as a curve becomes more complicated, extrapolation becomes subject to more error. Also, as the extent (or distance) of the extrapolation increases for a given curve, the accuracy decreases.

TRENDS

A function is said to be *nonincreasing* if the value of the dependent variable never grows any larger (or more positive) as the value of the independent

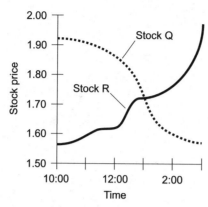

Fig. 2-14.   The price of Stock Q is nonincreasing versus time, and the price of Stock R is nondecreasing versus time.

variable increases. If the dependent variable in a function never gets any smaller (or more negative) as the value of the independent variable increases, the function is said to be *nondecreasing*.

The dashed curve in Fig. 2-14 shows the behavior of a hypothetical Stock Q, whose price never rises throughout the period under consideration. This function is nonincreasing. The solid curve shows a hypothetical Stock R, whose price never falls throughout the period. This function is nondecreasing.

Sometimes the terms *trending downward* and *trending upward* are used to describe graphs. These terms are subjective; different people might interpret them differently. Everyone would agree that Stock Q in Fig. 2-14 trends downward while Stock R trends upward. But a stock that rises and falls several times during a period might be harder to define in this respect.

CORRELATION

Graphs can show the extent of *correlation* between the values of two variables when the values are obtained from a finite number of experimental samples.

If, as the value of one variable generally increases, the value of the other generally increases too, the correlation is considered positive. If the opposite is true – the value of one variable increases as the other generally decreases – the correlation is considered negative. If the points are randomly scattered all over the graph, then the correlation is considered to be zero.

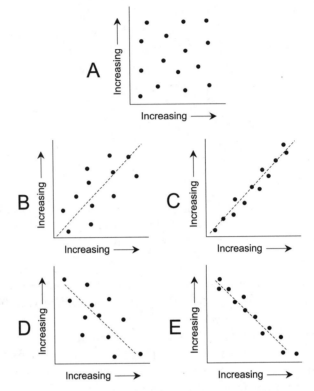

Fig. 2-15.   Point plots showing zero correlation (A), weak positive correlation (B), strong positive correlation (C), weak negative correlation (D), and strong negative correlation (E).

Figure 2-15 shows five examples of point sets. At A the correlation is zero. At B and C, the correlation is positive. At D and E, the correlation is negative.

PROBLEM 2-5
Suppose, as the value of the independent variable in a function changes, the value of the dependent variable does not change. This is called a *constant function*. Is its graph nonincreasing or nondecreasing?

SOLUTION 2-5
According to our definitions, the graph of a constant function is both nonincreasing and nondecreasing. Its value never increases, and it never decreases.

PROBLEM 2-6
Is there any type of function for which linear interpolation is perfectly accurate, that is, "fills in the gap" with zero error?

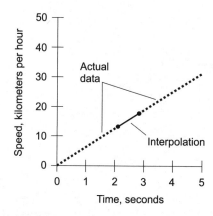

Fig. 2-16.   Illustration for Problem 2-6.

SOLUTION 2-6

Yes. If the graph of a function is known to be a straight line, then linear interpolation can be used to "fill in a gap" and the result will be free of error. An example is the speed of a car that accelerates at a known, constant rate. If its speed-versus-time graph appears as a perfectly straight line with a small gap, then linear interpolation can be used to determine the car's speed at points inside the gap, as shown in Fig. 2-16. In this graph, the heavy dashed line represents actual measured data, and the thinner solid line represents interpolated data.

Quiz

Refer to the text in this chapter if necessary. A good score is eight correct. Answers are in the back of the book.

1.   Suppose a graph is drawn showing temperature in degrees Celsius (°C) as a function of time for a 24-hour day. The temperature measurements are accurate to within 0.1 °C, and the readings are taken every 15 minutes. The day's highest temperature is +23.8 °C at 3:45 P.M. The day's lowest temperature is +13.5 °C at 5:30 A.M. A reasonable range for the temperature scale in this graph is
     (a) 0.0 °C to +100.0 °C
     (b) −100.0 °C to +100.0 °C
     (c) +15.0 °C to +20.0 °C
     (d) +10.0 °C to +30.0 °C

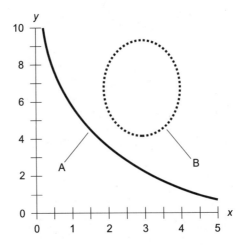

Fig. 2-17. Illustration for Quiz Questions 3 through 6.

2. In a constant function:
   (a) the value of the dependent variable constantly increases as the value of the independent variable increases
   (b) the value of the dependent variable constantly decreases as the value of the independent variable increases
   (c) the value of the independent variable does not change
   (d) the value of the dependent variable does not change

3. In Fig. 2-17, which of the curves represents y as a function of x?
   (a) Curve A only.
   (b) Curve B only.
   (c) Both curves A and B.
   (d) Neither curve A nor curve B.

4. In Fig. 2-17, which of the curves represents x as a function of y?
   (a) Curve A only.
   (b) Curve B only.
   (c) Both curves A and B.
   (d) Neither curve A nor curve B.

5. In Fig. 2-17, which of the curves represents a relation between x and y?
   (a) Curve A only.
   (b) Curve B only.
   (c) Both curves A and B.
   (d) Neither curve A nor curve B.

6. In Fig. 2-17, which of the curves can represent a constant function of one variable versus the other?
    (a) Curve A only.
    (b) Curve B only.
    (c) Both curves A and B.
    (d) Neither curve A nor curve B.

7. It is reasonable to suppose that the number of traffic accidents per year on a given section of highway is positively correlated with
    (a) the number of cars that use the highway
    (b) the number of sunny days in the year
    (c) the number of alternative routes available
    (d) the number of police officers that use the highway

8. Suppose a large data file is downloaded from the Internet. The speed of the data, in bits per second (bps), is plotted as a function of time in seconds. In this situation, time is
    (a) a dependent variable
    (b) an independent variable
    (c) a constant function
    (d) nondecreasing

9. Suppose the path of a hurricane is plotted as a curve on a map, with the latitude and longitude lines as graphic coordinates. A prediction of the future path of the storm can be attempted using
    (a) interpolation
    (b) extrapolation
    (c) curve fitting
    (d) correlation

10. Suppose the path of a hurricane is plotted as a series of points on a map, representing the position of the storm at 6-hour intervals. A continuous graph of the storm's path can be best approximated using
    (a) a bar graph
    (b) extrapolation
    (c) curve fitting
    (d) correlation

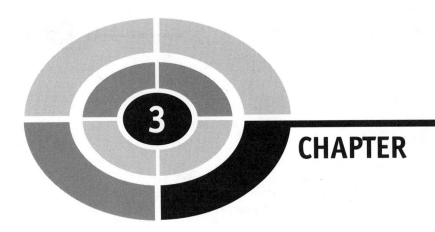

Extreme Numbers

A special technique is used to express, and to make calculations involving, gigantic and minuscule numbers. The scheme is like using a mathematical telescope or microscope. How many grains of sand are there on all the beaches in the world? What is the ratio of the diameter of a golf ball to the diameter of the moon? How many times brighter is the sun than a birthday candle?

Subscripts and Superscripts

Subscripts are used to modify the meanings of units, constants, and variables. A subscript is placed to the right of the main character (without spacing) and is set below the base line. *Superscripts* almost always represent *exponents* (the raising of a base quantity to a power). A superscript is placed to the right of the main character (without spacing) and is set above the base line.

EXAMPLES OF SUBSCRIPTS

Numeric subscripts are never italicized. Alphabetic subscripts sometimes are. Here are three examples of subscripted quantities:

Z_0 read "Z sub nought" or "Z sub zero":
        stands for characteristic impedance of a transmission line

R_{out} read "R sub out";
        stands for output resistance in an electronic circuit

y_n read "y sub n";
        represents a variable

Ordinary numbers are not normally modified with subscripts, so you are not likely to see expressions like this:

$$3_5$$
$$-9.7755_\pi$$
$$-16_x$$

Constants and variables can come in many "flavors." Some physical constants are assigned subscripts by convention. An example is m_e, representing the mass of an electron at rest. Sometimes, points in three-dimensional space are represented using ordered triples like (x_1, x_2, x_3) rather than (x, y, z). This subscripting scheme becomes especially convenient if you're talking about points in a higher-dimensional space, for example, $(x_1, x_2, x_3, \ldots, x_{26})$.

EXAMPLES OF SUPERSCRIPTS

Numeric superscripts are never italicized, but alphabetic superscripts usually are. Examples of superscripted quantities are:

2^3 read "two cubed";
        represents $2 \times 2 \times 2$

e^x read "e to the xth";
        represents the exponential function of x

$y^{1/2}$ read "y to the one-half";
        represents the square root of y

There is a significant difference between 2^3 and 2! There's also a big difference between the expression e that symbolizes the natural-logarithm base (approximately 2.71828) and e^x, which can represent e raised to a variable power, or which is sometimes used in place of the words "exponential function." Those little superscripts have a huge effect.

Power-of-10 Notation

Scientists and engineers express extreme numerical values using *power-of-10 notation*, also called *scientific notation*.

STANDARD FORM

A numeral in *standard power-of-10 notation* is written as follows:

$$m.n_1 n_2 n_3 \ldots n_p \times 10^z$$

where the dot (.) is a period, written on the base line (not a raised dot indicating multiplication), and is called the *radix point* or *decimal point*. The numeral m (to the left of the radix point) is a digit from the set {1, 2, 3, 4, 5, 6, 7, 8, 9}. The numerals n_1, n_2, n_3, and so on up to n_p (to the right of the radix point) are digits from the set {0, 1, 2, 3, 4, 5, 6, 7, 8, 9}. The value z, which is the power of 10, can be any integer: positive, negative, or zero. Here are some examples of numbers written in standard scientific notation:

$$2.56 \times 10^6$$
$$8.0773 \times 10^{-18}$$
$$1.000 \times 10^0$$

ALTERNATIVE FORM

In certain countries, and in some old books and papers, a slight variation on the above theme is used. The *alternative power-of-10 notation* requires that m always be equal to 0. When the above quantities are expressed this way, they appear as decimal fractions larger than 0 but less than 1, and the value of the exponent is increased by 1 compared with the standard form:

$$0.256 \times 10^7$$
$$0.80773 \times 10^{-17}$$
$$0.1000 \times 10^1$$

These expressions represent the same numbers as the previous three, even though they're written differently.

THE "TIMES SIGN"

The multiplication sign in a power-of-10 expression can be denoted in various ways. Most scientists in America use the cross symbol ($\times$), as in the examples shown above. But a small dot raised above the base line ($\cdot$) is sometimes used to represent multiplication in power-of-10 notation. When written that way, the above numbers look like this in the standard form:

$$2.56 \cdot 10^6$$

$$8.0773 \cdot 10^{-18}$$

$$1.000 \cdot 10^0$$

This small dot should not be confused with a radix point. The dot symbol is preferred when multiplication is required to express the dimensions of a physical unit. An example is the kilogram-meter per second squared, which is symbolized $kg \cdot m/s^2$ or $kg \cdot m \cdot s^{-2}$.

Another alternative multiplication symbol in power-of-10 notation is the asterisk ($*$). You will occasionally see numbers written like this:

$$2.56 * 10^6$$

$$8.0773 * 10^{-18}$$

$$1.000 * 10^0$$

PLAIN-TEXT EXPONENTS

Once in a while, you will have to express numbers in power-of-10 notation using plain, unformatted text. This is the case, for example, when transmitting information within the body of an e-mail message (rather than as an attachment). Some calculators and computers use this system. The letter E indicates that the quantity immediately following is a power of 10. In this format, the above quantities are written:

$$2.56E6$$

$$8.0773E-18$$

$$1.000E0$$

In an alternative format, the exponent is always written with two numerals, and always includes a plus sign or a minus sign, so the above expressions appear as:

$$2.56E+06$$

$$8.0773E-18$$

$$1.000E+00$$

Another alternative is to use an asterisk to indicate multiplication, and the symbol ^ to indicate a superscript, so the expressions look like this:

$$2.56 * 10\,\hat{}\,6$$

$$8.0773 * 10\,\hat{}-18$$

$$1.000 * 10\,\hat{}\,0$$

In all of these examples, the numerical values represented are identical. Respectively, if written out in full, they are:

$$2,560,000$$

$$0.0000000000000000080773$$

$$1.000$$

ORDERS OF MAGNITUDE

As you can see, power-of-10 notation makes it possible to easily write down numbers that denote unimaginably gigantic or tiny quantities. Consider the following:

$$2.55 \times 10^{45,589}$$

$$-9.8988 \times 10^{-7,654,321}$$

Imagine the task of writing either of these numbers out in ordinary decimal form! In the first case, you'd have to write the numerals 255, and then follow them with a string of 45,587 zeros. In the second case, you'd have to write a minus sign, then a numeral zero, then a radix point, then a string of 7,654,320 zeros, then the numerals 9, 8, 9, 8, and 8.

Now consider these two numbers:

$$2.55 \times 10^{45,592}$$

$$-9.8988 \times 10^{-7,654,318}$$

These look a lot like the first two, don't they? Look more closely. Both of these new numbers are 1000 times larger than the original two. You can tell by looking at the exponents. Both exponents are larger by 3. The number 45,592 is 3 more than 45,591, and the number −7,754,318 is 3 larger than −7,754,321. (Numbers grow larger in the mathematical sense as they become more positive or less negative.) The second pair of numbers is 3 *orders of magnitude* larger than the first pair of numbers. They look almost the same here, and they would look essentially identical if they were written out in full decimal form. But they are as different as a meter is from a kilometer.

The order-of-magnitude concept makes it possible to construct number lines, charts, and graphs with scales that cover huge spans of values. Three examples are shown in Fig. 3-1. Drawing A shows a number line spanning 3 orders of magnitude, from 1 to 1000. Illustration B shows a number line spanning 10 orders of magnitude, from 10^{-3} to 10^7. Illustration C shows a graph whose horizontal scale spans 10 orders of magnitude, from 10^{-3} to 10^7, and whose vertical scale extends from 0 to 10.

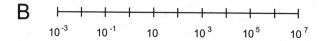

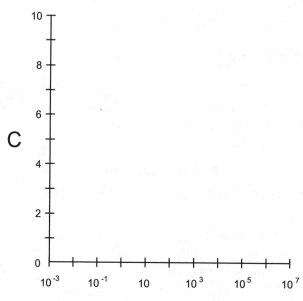

Fig. 3-1.   At A, a number line spanning three orders of magnitude. At B, a number line spanning 10 orders of magnitude. At C, a coordinate system whose horizontal scale spans 10 orders of magnitude, and whose vertical scale extends linearly from 0 to 10.

If you're astute, you'll notice that while the 0-to-10 scale is the easiest to directly envision, we can't really say how many orders of magnitude it covers. This is because, no matter how many times you cut a nonzero number to 1/10 its original size, you can never reach zero. In a certain sense, a linear scale (a scale where all the graduations are the same distance apart) ranging from 0 to any positive or negative value covers infinitely many orders of magnitude.

TO USE OR NOT TO USE

In printed literature, power-of-10 notation is generally used only when the power of 10 is large or small. If the exponent is between -2 and 2 inclusive, numbers are written out in plain decimal form as a rule. If the exponent is -3 or 3, numbers are sometimes written out, and are sometimes written in power-of-10 notation. If the exponent is -4 or smaller, or if it is 4 or larger, values are expressed in power-of-10 notation as a rule.

Some calculators, when set for power-of-10 notation, display all numbers that way, even those that normally shouldn't be. This can be confusing, especially when the power of 10 is zero and the calculator is set to display a lot of digits. Most people understand the expression 8.407 more easily than 8.407000000E+00, for example, even though they represent the same number.

PREFIX MULTIPLIERS

Special verbal prefixes, known as *prefix multipliers*, are commonly used by physicists and engineers to express orders of magnitude. Table 3-1 shows the prefix multipliers used for factors ranging from 10^{-24} to 10^{24}.

PROBLEM 3-1
By how many orders of magnitude does a *terahertz* differ from a *megahertz*? (The *hertz* is a unit of frequency, equivalent to a cycle per second.)

SOLUTION 3-1
Refer to Table 3-1. A terahertz represents 10^{12} hertz, and a megahertz represents 10^6 hertz. The exponents differ by 6. Therefore, a terahertz differs from a megahertz by 6 orders of magnitude.

PROBLEM 3-2
What, if anything, is wrong with the number 344.22×10^7 as an expression in power-of-10 notation?

Table 3-1   Prefix multipliers and their abbreviations.

Designator	Symbol	Multiplier
yocto-	y	10^{-24}
zepto-	z	10^{-21}
atto-	a	10^{-18}
femto-	f	10^{-15}
pico-	p	10^{-12}
nano-	n	10^{-9}
micro-	μ or mm	10^{-6}
milli-	m	10^{-3}
centi-	c	10^{-2}
deci-	d	10^{-1}
(none)	–	10^{0}
deka-	da or D	10^{1}
hecto-	h	10^{2}
kilo-	K or k	10^{3}
mega-	M	10^{6}
giga-	G	10^{9}
tera-	T	10^{12}
peta-	P	10^{15}
exa-	E	10^{18}
zetta-	Z	10^{21}
yotta-	Y	10^{24}

SOLUTION 3-2

This is a legitimate number, but it is not written in the correct format for scientific notation. The number to the left of the multiplication symbol should be at least 1, but smaller than 10. To convert the number to the proper format, first divide the portion to the left of the "times sign" by 100, so it becomes 3.4422. Then multiply the portion to the right of the "times sign" by 100, increasing the exponent by 2 so it becomes 10^9. This produces the same number in correct power-of-10 format: 3.4422×10^9.

In Action

Let's see how power-of-10 notation works when we want to do simple arithmetic using extreme numbers.

ADDITION

Addition is best done by converting the numbers to ordinary decimal form before performing the operation. Scientific notation doesn't work very well with addition unless both numbers happen to be expressed in the same power of 10. Here are a couple of examples:

$$(3.045 \times 10^5) + (6.853 \times 10^6)$$

$$= 304{,}500 + 6{,}853{,}000$$

$$= 7{,}157{,}500$$

$$= 7.1575 \times 10^6$$

$$(3.045 \times 10^{-4}) + (6.853 \times 10^{-7})$$

$$= 0.0003045 + 0.0000006853$$

$$= 0.0003051853$$

$$= 3.051853 \times 10^{-4}$$

$$(3.045 \times 10^5) + (6.853 \times 10^{-7})$$

$$= 304{,}500 + 0.0000006853$$

$$= 304{,}500.0000006853$$

$$= 3.045000000006853 \times 10^5$$

SUBTRACTION

Subtraction follows the same basic rules as addition:

$$(3.045 \times 10^5) - (6.853 \times 10^6)$$
$$= 304{,}500 - 6{,}853{,}000$$
$$= -6{,}548{,}500$$
$$= -6.548500 \times 10^6$$

$$(3.045 \times 10^{-4}) - (6.853 \times 10^{-7})$$
$$= 0.0003045 - 0.0000006853$$
$$= 0.0003038147$$
$$= 3.038147 \times 10^{-4}$$

$$(3.045 \times 10^5) - (6.853 \times 10^{-7})$$
$$= 304{,}500 - 0.0000006853$$
$$= 304{,}499.9999993147$$
$$= 3.044999999993147 \times 10^5$$

MULTIPLICATION

When numbers are multiplied in power-of-10 notation, the coefficients (the numbers to the left of the multiplication symbol) are multiplied by each other. Then the exponents are added. Finally, the product is reduced to standard form. Here are three examples, using the same three number pairs as before:

$$(3.045 \times 10^5) \times (6.853 \times 10^6)$$
$$= (3.045 \times 6.853) \times (10^5 \times 10^6)$$
$$= 20.867385 \times 10^{(5+6)}$$
$$= 20.867385 \times 10^{11}$$
$$= 2.0867385 \times 10^{12}$$

$$(3.045 \times 10^{-4}) \times (6.853 \times 10^{-7})$$
$$= (3.045 \times 6.853) \times (10^{-4} \times 10^{-7})$$
$$= 20.867385 \times 10^{[-4+(-7)]}$$
$$= 20.867385 \times 10^{-11}$$
$$= 2.0867385 \times 10^{-10}$$

$$(3.045 \times 10^{5}) \times (6.853 \times 10^{-7})$$
$$= (3.045 \times 6.853) \times (10^{5} \times 10^{-7})$$
$$= 20.867385 \times 10^{(5-7)}$$
$$= 20.867385 \times 10^{-2}$$
$$= 2.0867385 \times 10^{-1}$$
$$= 0.20867385$$

This last number is written out in plain decimal form because the exponent is between -2 and 2 inclusive.

DIVISION

When numbers are divided in power-of-10 notation, the coefficients are divided by each other. Then the exponents are subtracted. Finally, the quotient is reduced to standard form. Here's how division works, with the same three number pairs we've been using:

$$(3.045 \times 10^{5})/(6.853 \times 10^{6})$$
$$= (3.045/6.853) \times (10^{5}/10^{6})$$
$$\approx 0.444331 \times 10^{(5-6)}$$
$$= 0.444331 \times 10^{-1}$$
$$= 0.0444331$$

$$(3.045 \times 10^{-4})/(6.853 \times 10^{-7})$$
$$= (3.045/6.853) \times (10^{-4}/10^{-7})$$
$$\approx 0.444331 \times 10^{[-4-(-7)]}$$
$$= 0.444331 \times 10^{3}$$
$$= 4.44331 \times 10^{2}$$
$$= 444.331$$

$$(3.045 \times 10^5)/(6.853 \times 10^{-7})$$

$$= (3.045/6.853) \times (10^5/10^{-7})$$

$$\approx 0.444331 \times 10^{[5-(-7)]}$$

$$= 0.444331 \times 10^{12}$$

$$= 4.44331 \times 10^{11}$$

Note the squiggly equal signs ($\approx$) in the above equations. This symbol means "is approximately equal to." The numbers here don't divide out neatly, so the decimal portions are approximated.

EXPONENTIATION

When a number is raised to a power in scientific notation, both the coefficient and the power of 10 itself must be raised to that power, and the result multiplied. Consider this example:

$$(4.33 \times 10^5)^3$$

$$= (4.33)^3 \times (10^5)^3$$

$$= 81.182737 \times 10^{(5 \times 3)}$$

$$= 81.182737 \times 10^{15}$$

$$= 8.1182727 \times 10^{16}$$

If you are a mathematical purist, you will notice gratuitous parentheses in the second and third lines here. These are included to minimize the chance for confusion. It is more important that the result of a calculation be correct than that the expression be mathematically lean.

Let's consider another example, in which the exponent is negative:

$$(5.27 \times 10^{-4})^2$$

$$= (5.27)^2 \times (10^{-4})^2$$

$$= 27.7729 \times 10^{(-4 \times 2)}$$

$$= 27.7729 \times 10^{-8}$$

$$= 2.77729 \times 10^{-7}$$

TAKING ROOTS

To find the root of a number in power-of-10 notation, think of the root as a fractional exponent. The square root is equivalent to the 1/2 power. The cube root is the same thing as the 1/3 power. In general, the nth root of a number (where n is a positive integer) is the same thing as the $1/n$ power. When roots are regarded this way, it is easy to multiply things out in exactly the same way as is done with whole-number exponents. Here is an example:

$$(5.27 \times 10^{-4})^{1/2}$$
$$= (5.27)^{1/2} \times (10^{-4})^{1/2}$$
$$\approx 2.2956 \times 10^{[-4 \times (1/2)]}$$
$$= 2.2956 \times 10^{-2}$$
$$= 0.02956$$

Note the squiggly equal sign in the third line. The square root of 5.27 is an irrational number, and the best we can do is to approximate its decimal expansion because it cannot be written out in full. Also, the exponent in the final result is within the limits for which we can write the number out in plain decimal form, so power-of-10 notation is not needed to express the end product here.

PROBLEM 3-3
State the rule for multiplication in scientific notation, using the variables u and v to represent the coefficients and the variables m and n to represent the exponents.

SOLUTION 3-3
Let u and v be real numbers greater than or equal to 1 but less than 10, and let m and n be integers. Then the following equation holds true:

$$(u \times 10^{m})(v \times 10^{n}) = uv \times 10^{(m+n)}$$

PROBLEM 3-4
State the rule for division in scientific notation, using the variables u and v to represent the coefficients and the variables m and n to represent the exponents.

SOLUTION 3-4
Let u and v be real numbers greater than or equal to 1 but less than 10, and let m and n be integers. Then the following equation holds true:

$$(u \times 10^{m})/(v \times 10^{n}) = u/v \times 10^{(m-n)}$$

PROBLEM 3-5

State the rule for exponentiation in scientific notation, using the variable u to represent the coefficient, and the variables m and n to represent the exponents.

SOLUTION 3-5

Let u be a real number greater than or equal to 1 but less than 10, and let m and n be integers. Then the following equation holds true:

$$(u \times 10^m)^n = u^n \times 10^{(mn)}$$

Exponents m and n can be real numbers in general. They need not be restricted to integer values. But that subject is too advanced for this chapter. We'll be dealing with them in Chapter 14, when we get to logarithms and exponential functions.

Approximation and Precedence

Numbers in the real world are not always exact. This is especially true in observational science and in engineering. Often, we must approximate. There are two ways of doing this: *truncation* (simpler but less accurate) and *rounding* (a little more difficult, but more accurate).

TRUNCATION

The process of truncation involves the deletion of all the numerals to the right of a certain point in the decimal part of an expression. Some electronic calculators use truncation to fit numbers within their displays. For example, the number 3.830175692803 can be shortened in steps as follows:

3.830175692803
3.83017569280
3.8301756928
3.830175692
3.83017569
3.8301756
3.830175
3.83017
3.83
3.8
3

ROUNDING

Rounding is the preferred method of rendering numbers in shortened form. In this process, when a given digit (call it r) is deleted at the right-hand extreme of an expression, the digit q to its left (which becomes the new r after the old r is deleted) is not changed if $0 \leq r \leq 4$. If $5 \leq r \leq 9$, then q is increased by 1 ("rounded up"). Most electronic calculators use rounding rather than truncation. If rounding is used, the number 3.830175692803 can be shortened in steps as follows:

3.830175692803

3.83017569280

3.8301756928

3.830175693

3.83017569

3.8301757

3.830176

3.83018

3.8302

3.830

3.83

3.8

4

PRECEDENCE

In scientific notation, as in regular arithmetic, operations should be performed in a certain order when they appear together in an expression. When more than one type of operation appears, and if you need to simplify the expression, the operations should be done in the following order:

- Simplify all expressions within parentheses, brackets, and braces from the inside out.
- Perform all exponential operations, proceeding from left to right.
- Perform all products and quotients, proceeding from left to right.
- Perform all sums and differences, proceeding from left to right.

Here are two examples of expressions simplified according to the above rules of precedence. Note that the order of the numerals and operations is the same in each case, but the groupings differ.

$$3.4 \times 10^5 + 5.0 \times 10^{-4}$$

$$= (3.4 \times 10^5) + (5.0 \times 10^{-4})$$

$$= (340,000 + 0.0005)$$

$$= 3.400005 \times 10^5$$

$$3.4 \times (10^5 + 5.0) \times 10^4$$

$$= 3.4 \times 100,005 \times 10,000$$

$$= 3,400,170,000$$

$$= 3.40017 \times 10^9$$

Suppose you're given a complicated expression and there are no parentheses, brackets, or braces in it. This is not ambiguous if the above-mentioned rules are followed. Consider this example:

$$z = -3x^3 + 4x^2y - 12xy^2 - 5y^3$$

If this were written with parentheses, brackets, and braces to emphasize the rules of precedence, it would look like this:

$$z = [-3(x^3)] + \{4[(x^2)y]\} - \{12[x(y^2)]\} - [5(y^3)]$$

Because we have agreed on the rules of precedence, we can do without the parentheses, brackets, and braces. Nevertheless, if there is any doubt about a crucial equation, you're better off to use a couple of unnecessary parentheses than to make a costly mistake.

PROBLEM 3-6
What is the value of $2 + 3 \times 4 + 5$?

SOLUTION 3-6
First, perform the multiplication operation, obtaining the expression $2 + 12 + 5$. Then add the numbers, obtaining the final value 19. Therefore:

$$2 + 3 \times 4 + 5 = 19$$

Significant Figures

The number of significant figures, also called significant digits, in a power-of-10 number is important in real-world mathematics, because it indicates the degree of accuracy to which we know a particular value.

MULTIPLICATION, DIVISION, AND EXPONENTIATION

When multiplication, division, or exponentiation is done using power-of-10 notation, the number of significant figures in the result cannot legitimately be greater than the number of significant figures in the least-exact expression. You might wonder why, in some of the above examples, we come up with answers that have more digits than any of the numbers in the original problem. In pure mathematics, where all numbers are assumed exact, this is not an issue. But in real life, and especially in experimental science and engineering, things are not so clean-cut.

Consider the two numbers $x = 2.453 \times 10^4$ and $y = 7.2 \times 10^7$. The following is a perfectly valid statement in pure mathematics:

$$xy = 2.453 \times 10^4 \times 7.2 \times 10^7$$

$$= 2.453 \times 7.2 \times 10^{11}$$

$$= 17.6616 \times 10^{11}$$

$$= 1.76616 \times 10^{12}$$

But if x and y represent measured quantities, as in experimental physics for example, the above statement needs qualification. We must pay close attention to how much accuracy we claim.

HOW ACCURATE ARE WE?

When you see a product or quotient containing a bunch of numbers in scientific notation, count the number of single digits in the coefficients of each number. Then take the smallest number of digits. That's the number of significant figures you can claim in the final answer or solution.

In the above example, there are four single digits in the coefficient of x, and two single digits in the coefficient of y. So you must round off the answer, which appears to contain six significant figures, to two. It is important to use rounding, and not truncation. You should conclude that:

$$xy = 2.453 \times 10^4 \times 7.2 \times 10^7$$
$$= 1.8 \times 10^{12}$$

In situations of this sort, if you insist on being rigorous, you should use squiggly equal signs throughout, because you are always dealing with approximate values. But most folks are content to use ordinary equals signs. It is universally understood that physical measurements are inherently inexact, and writing squiggly lines can get tiresome.

Suppose you want to find the quotient x/y instead of the product xy? Proceed as follows:

$$x/y = (2.453 \times 10^4)/(7.2 \times 10^7)$$
$$= (2.453/7.2) \times 10^{-3}$$
$$= 0.3406944444\ldots \times 10^{-3}$$
$$= 3.406944444\ldots \times 10^{-4}$$
$$= 3.4 \times 10^{-4}$$

WHAT ABOUT ZEROS?

Sometimes, when you make a calculation, you'll get an answer that lands on a neat, seemingly whole-number value. Consider $x = 1.41421$ and $y = 1.41422$. Both of these have six significant figures. The product, taking significant figures into account, is:

$$xy = 1.41421 \times 1.41422$$
$$= 2.0000040662$$
$$= 2.00000$$

This looks like it's exactly equal to 2. In pure mathematics, $2.00000 = 2$. But in practical math, those five zeros indicate how near the exact number 2 we believe the resultant to be. We know the answer is very close to a mathematician's idea of the number 2, but there is an uncertainty of up to plus-or-minus 0.000005 (written ± 0.000005). When we claim a certain number of significant figures, zero is as important as any of the other nine digits in the decimal number system.

ADDITION AND SUBTRACTION

When measured quantities are added or subtracted, determining the number of significant figures can involve subjective judgment. The best procedure is to expand all the values out to their plain decimal form (if possible), make the calculation as if you were a pure mathematician, and then, at the end of the process, decide how many significant figures you can reasonably claim.

In some cases, the outcome of determining significant figures in a sum or difference is similar to what happens with multiplication or division. Take, for example, the sum $x + y$, where $x = 3.778800 \times 10^{-6}$ and $y = 9.22 \times 10^{-7}$. This calculation proceeds as follows:

$$x = 0.000003778800$$
$$y = 0.000000922$$
$$x + y = 0.0000047008$$
$$= 4.7008 \times 10^{-6}$$
$$= 4.70 \times 10^{-6}$$

In other instances, one of the values in a sum or difference is insignificant with respect to the other. Let's say that $x = 3.778800 \times 10^4$, while $y = 9.22 \times 10^{-7}$. The process of finding the sum goes like this:

$$x = 37,788.00$$
$$y = 0.000000922$$
$$x + y = 37,788.000000922$$
$$= 3.7788000000922 \times 10^4$$

In this case, y is so much smaller than x that it doesn't significantly affect the value of the sum. Here, it is best to regard y, in relation to x or to the sum $x + y$, as the equivalent of a gnat compared with a watermelon. If a gnat lands on a watermelon, the total weight does not appreciably change, nor does the presence or absence of the gnat have any effect on the accuracy of the scales. We can conclude that the "sum" here is the same as the larger number. The value y is akin to a nuisance or a negligible error:

$$x + y = 3.778800 \times 10^4$$

PROBLEM 3-7

What is the product of 1.001×10^5 and 9.9×10^{-6}, taking significant figures into account?

SOLUTION 3-7
Multiply the coefficients and the powers of 10 separately:

$$(1.001 \times 10^5)(9.9 \times 10^{-6})$$

$$= (1.001 \times 9.9) \times (10^5 \times 10^{-6})$$

$$= 9.9099 \times 10^{-1}$$

$$= 0.99099$$

We must round this to two significant figures, because that is the most we can legitimately claim. The expression does not have to be written out in power-of-10 form. Therefore:

$$(1.001 \times 10^5)(9.9 \times 10^{-6}) = 0.99$$

Quiz

Refer to the text in this chapter if necessary. A good score is eight correct. Answers are in the back of the book.

1. Two numbers differ in size by exactly 4 orders of magnitude. This is a factor of:
   (a) 4
   (b) 16
   (c) 256
   (d) 10,000

2. Suppose the population of a certain city is found to be 226,496 people. What is this expressed to three significant figures in power-of-10 notation?
   (a) 2.26×10^5
   (b) 2.27×10^5
   (c) 227,000
   (d) 2.26E+06

3. Suppose we invent a new unit called the *widget* (symbol: Wi). What should we call a quantity of 10^{-12} Wi?
   (a) One milliwidget (1 mWi).
   (b) One nanowidget (1 nWi).
   (c) One picowidget (1 pWi).
   (d) One kilowidget (1 kWi).

4. What is the value of $5 \times 3^2 - 5$?
   (a) 20
   (b) 40
   (c) 220
   (d) There is no way to tell; this is an ambiguous expression.

5. What is $8.899 \times 10^5 - 2.02 \times 10^{-12}$, taking significant figures into account?
   (a) 2.02×10^{-12}
   (b) 9×10^5
   (c) 6.88×10^5
   (d) 8.899×10^5

6. What is the product of 8.72×10^5 and 6.554×10^{-5}, taking significant figures into account?
   (a) 57.2
   (b) 57.15
   (c) 57.151
   (d) 57.1509

7. Suppose a mathematical expression contains addition, multiplication, and exponentiation, all without any parentheses, brackets, or braces. Which operation should be performed last?
   (a) The exponentiation.
   (b) The multiplication.
   (c) The addition.
   (d) It doesn't matter.

8. Suppose we measure a quantity and get 5.53×10^4 units, accurate to three significant figures. Within what range of whole-number units can we say the actual value is?
   (a) 55,299 units to 56,301 units.
   (b) 55,290 units to 56,310 units.
   (c) 55,250 units to 55,349 units.
   (d) 55,200 units to 55,299 units.

9. What is the proper power-of-10 way of writing 416,803?
   (a) 416.803×10^3
   (b) 41.6803E+04
   (c) 0.0416803×10^7
   (d) None of the above.

10. What is the order in which the operations should be performed in the following expression?

$$S = x(y + z) - w^2$$

   (a) Addition, then subtraction, then exponentiation, and finally multiplication.
   (b) Addition, then exponentiation, then multiplication, and finally subtraction.
   (c) Multiplication, then exponentiation, then subtraction, and finally addition.
   (d) Subtraction, multiplication, exponentiation, and finally addition.

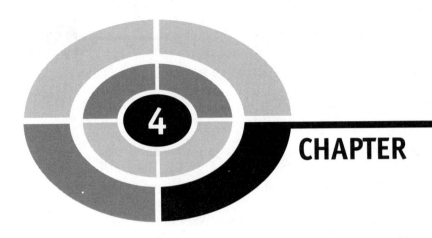

How Things Are Measured

Numbers by themselves are abstract. *Units* make numbers concrete. We can say that a quantity is this wide, is that heavy, lasts this long, or is that hot; it extends for this many millimeters, weighs that many pounds, lasts this many hours, or is heated to that many degrees Celsius. Some of the material in this chapter might at first seem too scientific to be considered "everyday math." But when it comes to math, it's always important to know precisely what you mean when you say something. After you've gone through this material, nobody will be able to legitimately say that you don't know what you're talking about when you quote units of measurement.

CHAPTER

4

Systems of Units

There are various systems of physical units. The *meter/kilogram/second (mks) system*, also called the *metric system* and which is essentially the same as the *International System*, is favored in most of the developed world. The *centimeter/gram/second (cgs) system* is less often used. The *foot/pound/second (fps) system*, also called the *English system*, is popular among nonscientists in the United States. Each system has several fundamental, or *base*, units from which all the others are derived.

THE INTERNATIONAL SYSTEM (SI)

The International System is abbreviated SI (for the words *Système International* in French). This scheme in its earlier form, mks, has existed since the 1800s, but was more recently defined by the General Conference on Weights and Measures.

The *base units* in SI quantify *displacement*, *mass*, *time*, *temperature*, *electric current*, *brightness of light*, and *amount of matter*. Respectively, the units in SI are known as the *meter*, the *kilogram*, the *second*, the *Kelvin* (or *degree Kelvin*), the *ampere*, the *candela*, and the *mole*.

THE CGS SYSTEM

In the centimeter/gram/second (cgs) system, the base units are the *centimeter* (exactly 0.01 meter), the *gram* (exactly 0.001 kilogram), the second (identical to the SI second), the *degree Celsius* (approximately the number of Kelvins minus 273.15), the ampere (identical to the SI ampere), the candela (identical to the SI candela), and the mole (identical to the SI mole).

THE ENGLISH SYSTEM

In the English or fps system, the base units are the *foot* (approximately 30.5 centimeters), the *pound* (equivalent to about 2.2 kilograms in the gravitational field at the earth's surface), and the second (identical to the SI second). Other units include the *degree Fahrenheit* (where water freezes at 32 degrees and boils at 212 degrees at standard sea-level atmospheric pressure), the ampere (identical to the SI ampere), the candela (identical to the SI candela), and the mole (identical to the SI mole).

Base Units in SI

In all systems of measurement, the base units are those from which all the others can be derived. Base units represent some of the most elementary properties or phenomena we observe in nature.

THE METER

The fundamental unit of distance, length, linear dimension, or displacement (all these terms mean essentially the same thing) is the meter, symbolized by the non-italicized, lowercase English letter m.

The original definition of the meter was based on the notion of dividing the shortest possible surface route between the north pole and the equator of the earth, as it would be measured if it passed through Paris, France (Fig. 4-1), into 10 million (10^7) identical units. Mountains, bodies of water, and other barriers were ignored; the earth was imagined to be a perfectly round, smooth ball. The circumference of the earth was therefore determined to be 40 million (4.000×10^7) m, give or take a little depending on which great circle around the globe one happened to choose.

Nowadays, the meter is defined more precisely as the distance a beam of light travels through a perfect vacuum in 1/299,792,458, or $3.33564095 \times 10^{-9}$, of a second. This is approximately the length of an adult's full stride when walking at a brisk pace, and is a little longer than an English yard.

Various units smaller or larger than the meter are often employed to measure or define displacement. A *millimeter* (mm) is 0.001 m. A *micrometer*

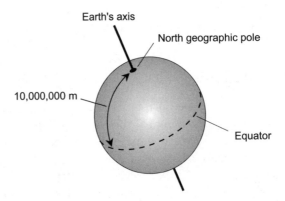

Fig. 4-1.   There are about 10,000,000 meters between the earth's north pole and the equator.

(μm or μ) is 0.000001 m or 10^{-6} m. Sometimes the micrometer is called the *micron*, but that term is technically out-of-date. A *nanometer* (nm) is 10^{-9} m. A *kilometer* (km) is 1000 m.

THE KILOGRAM

The base SI unit of mass is the kilogram, symbolized by the lowercase, non-italicized pair of English letters kg.

Originally, the kilogram was defined as the mass of a cube of pure liquid water measuring exactly 0.1 m on each edge (Fig. 4-2). That is still an excellent definition, but these days, scientists have come up with something more absolute. A kilogram is the mass of a sample of platinum–iridium alloy that is kept at the International Bureau of Weights and Measures. This object is called the *international prototype of the kilogram.*

It's important to realize that mass is not the same thing as *weight*. A mass of 1 kg maintains this same mass no matter where it is located. That standard platinum–iridium ingot would have a mass of 1 kg on the moon, on Mars, or in intergalactic space. Weight, in contrast, is a force exerted by gravitation on a given mass. On the surface of the earth, a 1-kg mass happens to weigh about 2.2 pounds.

Various units smaller or larger than the kilogram are used to measure or define mass. A *gram* (g) is 0.001 kg. A *milligram* (mg) is 0.000001 kg or 10^{-6} kg. A *microgram* (μg) is 10^{-9} kg. A *nanogram* (ng) is 10^{-12} g. A *metric ton* is 1000 kg.

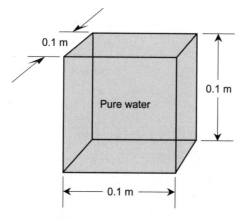

Fig. 4-2.   Originally, the kilogram was defined as the mass of 0.001 cubic meter of pure liquid water.

THE SECOND

The SI unit of time is the second, symbolized by the lowercase, non-italic English letter s. Sometimes it is abbreviated using its first three letters, sec.

The second was originally defined as exactly 1/60 of a minute, which is exactly 1/60 of an hour, which in turn is exactly 1/24 of a *mean solar day*. (The *solar day* is defined as the length of time it takes for the sun to pass from due south to due south again as observed from the earth's northern hemisphere; the mean solar day is the average length of all the solar days in a year.) A second was thought of as 1/86,400 of a mean solar day, and that is still an excellent definition (Fig. 4-3). But more recently, with the availability of atomic time standards, 1 s has been formally defined as the amount of time taken for a certain isotope of elemental cesium to oscillate through 9,192,631,770 (9.192631770×10^9) complete cycles.

A second happens to be the time it takes for a ray of light to travel 2.99792458×10^8 m through space. This is about three-quarters of the way to the moon. You might have heard of the moon being a little more than one *light-second* away from earth. If you are old enough to remember the conversations earth-based personnel carried on with Apollo astronauts as the space travelers walked around on the moon, you will recall the delay between comments or questions from earthlings and replies from the moonwalkers.

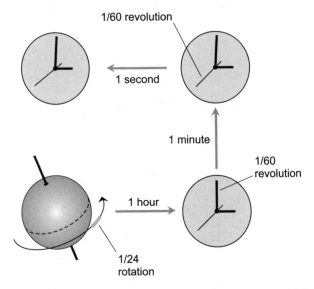

Fig. 4-3.   Originally, the second was defined as (1/60)(1/60)(1/24), or 1/86,400, of a mean solar day.

The astronauts were not hesitating; it took more than 2 s for radio signals to make a round trip between the earth and moon.

Units smaller than the second are often employed to measure or define time. A *millisecond* (ms) is 0.001 s; a *microsecond* (μs) is 0.000001 s or 10^{-6} s; a *nanosecond* (ns) is 10^{-9} s.

THE KELVIN

The SI unit of temperature is the Kelvin, symbolized K or °K (uppercase and non-italicized). It is a measure of how much heat exists relative to *absolute zero*, which represents the absence of all heat and which is therefore the coldest possible temperature. A temperature of 0 °K represents absolute zero. There are never any negative temperature readings on the Kelvin scale.

Formally, the Kelvin is defined as a temperature increment (an increase or decrease) of 1/273.16, or 0.0036609, part of the thermodynamic temperature of the *triple point* of pure water. The triple point of water is the temperature and pressure at which it can exist as vapor, liquid, and ice in equilibrium. It's almost, but not exactly, the same as the freezing point. Pure water at sea level freezes (or melts) at +273.15°K, and boils (or condenses) at +373.15°K.

THE AMPERE

The ampere, symbolized by the non-italic, uppercase English letter A (or abbreviated as amp), is the unit of electric current. A flow of approximately 6.241506×10^{18} electrons per second, past a given fixed point in an electrical conductor, represents a current of 1 A.

Various units smaller than the ampere are often employed to measure or define current. A *milliampere* (mA) is 0.001 A, or a flow of 6.241506×10^{15} electrons per second past a given fixed point. A *microampere* (μA) is 0.000001 A or 10^{-6} A, or a flow of 6.241506×10^{12} electrons per second. A *nanoampere* (nA) is 10^{-9} A; it is the smallest unit of electric current you are likely to hear about. It represents a flow of 6.241506×10^{9} electrons per second past a given fixed point.

The most rigorous definition of the ampere is highly theoretical: 1 A is the amount of constant charge-carrier flow through two straight, parallel, infinitely thin, perfectly conducting wires, placed 1 m apart in a vacuum, that results in a certain amount of force (2×10^{-7} *newton*) per meter of distance. The newton will be defined shortly.

THE CANDELA

The candela, symbolized by the non-italicized, lowercase pair of English letters cd, is the unit of luminous intensity. It is equivalent to 1/683 of a watt of radiant energy, emitted at a frequency of 5.40×10^{14} hertz (cycles per second), in a solid angle of 1 *steradian*. The watt and the steradian will be defined shortly.

There is a simpler, although crude, definition of the candela: 1 cd is roughly the amount of light emitted by an ordinary candle. (That's why brightness used to be expressed in units called *candlepower*.) Yet another definition tells it this way: 1 cd represents the radiation from a surface area of 1.667×10^{-6} square meters of a perfectly-radiating object called a *blackbody*, at the solidification temperature of pure platinum.

THE MOLE

The mole, symbolized or abbreviated by the non-italicized, lowercase English letters mol, is the standard unit of material quantity. It is also known as *Avogadro's number*, and is a huge number, approximately equal to 6.022169×10^{23}. This is the number of atoms in exactly 12 g (0.012 kg) of carbon 12, the most common isotope of elemental carbon with six protons and six neutrons in the nucleus.

The mole arises naturally in the physical world, especially in chemistry. It is one of those numbers for which the cosmos seems to have reserved a special place.

You'll sometimes hear about units called *millimoles* (mmol) and *kilomoles* (kmol). A quantity of 1 mmol represents 0.001 mol or 6.022169×10^{20}; a quantity of 1 kmol represents 1000 mol or 6.022169×10^{26}. The mole is an example of a *dimensionless unit* because it does not express any particular physical phenomenon. It's just a certain standardized number, like a dozen (12) or a gross (144).

FOR FURTHER INFORMATION

The history of the International System of Units is interesting. The definitions of the base units have gone through a lot of evolution! If you want to know more, visit the following Web site (valid at the time of this writing):

http://physics.nist.gov/cuu/Units/history.html

If this URL (uniform resource locator) does not bring up the right information, or if the link is broken, bring up one of the popular search engines such as Google (http://www.google.com) or Yahoo (http://www.yahoo.com), and input "International System of Units" using the phrase-search feature.

A NOTE ABOUT SYMBOLOGY

Until now, we've been rigorous about mentioning that symbols and abbreviations consist of lowercase or uppercase, non-italicized letters or strings of letters. That's important, because if this distinction is not made, especially relating to the use of italics, the symbols or abbreviations for physical units can be confused with the constants, variables, or coefficients that appear in equations.

When a letter is italicized, it almost always represents a constant, a variable, or a coefficient. When it is non-italicized, it often represents a physical unit or a prefix multiplier. A good example is s, which represents second, versus s, which is often used to represent linear dimension or displacement. Another example is m, representing meter or meters, as compared with m, which is used to denote the slope of a line in a graph.

From now on, we won't belabor this issue every time a unit symbol or abbreviation comes up. But don't forget it. Like the business about significant figures, this seemingly trivial thing can matter a lot!

PROBLEM 4-1
Suppose a pan of water is heated uniformly at the steady rate of 0.001 °K per second from 290 °K to 320 °K. (Water is a liquid at these temperatures on the earth's surface.) Draw a graph of this situation, where time is the independent variable and is plotted on the horizontal axis, and temperature is the dependent variable and is plotted on the vertical axis for values between 311 °K and 312 °K only. Optimize the time scale.

SOLUTION 4-1
The optimized graph is shown in Fig. 4-4. Note that it's a straight line, indicating that the temperature of the water rises at a constant rate. Because a change of only 0.001 °K takes place every second, it takes 1000 seconds for the temperature to rise the 1 °K from 311 °K to 312 °K. The time scale is therefore graduated in relative terms from 0 to 1000.

PROBLEM 4-2
Draw a "zoomed-in" version of the graph of Fig. 4-4, showing only the temperature range from 311.30 °K to 311.40 °K. Use a relative time scale, starting at 0. Again, optimize the time scale.

SOLUTION 4-2

See Fig. 4-5. The line is in the same position on the coordinate grid as the one shown in the previous example. The scales, however, have both been "magnified" or "zoomed-into" by a factor of 10.

PROBLEM 4-3

How many electrons flow past a given point in 3.00 s, given an electrical current of 2.00 A? Express the answer to three significant figures.

SOLUTION 4-3

From the definition of the ampere, we know that 1.00 A of current represents the flow of 6.241506×10^{18} electrons per second past a point per second

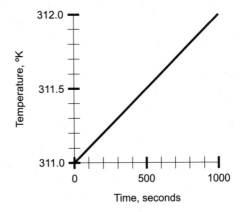

Fig. 4-4. Illustration for Problem 4-1.

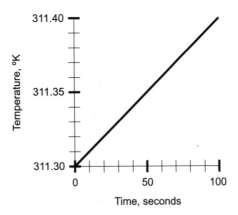

Fig. 4-5. Illustration for Problem 4-2.

of time. Therefore, a current of 2.00 A represents twice this many electrons, or 1.2483012×10^{19}, flowing past the point each second. In 3.00 seconds, three times that many electrons pass the point. That's 3.7449036×10^{19} electrons. Rounding off to three significant figures, we get the answer: 3.74×10^{19} electrons pass the point in 3.00 s.

Other Units in SI

The preceding seven units can be combined to generate many other units. Sometimes, the so-called *derived units* are expressed in terms of the base units, although such expressions can be confusing (for example, seconds cubed or kilograms to the minus-one power). If you ever come across combinations of units that seem nonsensical, don't be alarmed. You are looking at a derived unit that has been put down in terms of base units.

THE RADIAN

The standard unit of *plane angular measure* is the *radian* (rad). It is the angle subtended by an arc on a circle, whose length, as measured on the circle, is equal to the radius of the circle as measured on a flat geometric plane containing the circle. Imagine taking a string and running it out from the center of a circle to some point on the edge, and then laying that string down around the periphery of the circle. The resulting angle is 1 rad. Another definition goes like this: one radian is the angle between the two straight edges of a slice of pie whose straight and curved edges all have the same length r (Fig. 4-6). It is equal to about 57.2958 *angular degrees*.

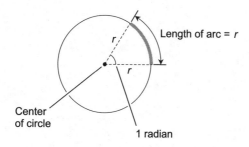

Fig. 4-6.   One radian is the angle at the apex of a slice of pie whose straight and curved edges all have the same length r.

THE ANGULAR DEGREE

The angular degree, symbolized by a little elevated circle (°) or by the three-letter abbreviation deg, is equal to 1/360 of a complete circle. The history of the degree is uncertain, although one theory says that ancient mathematicians chose it because it represents approximately the number of days in the year. One angular degree is equal to approximately 0.0174533 radians.

THE STERADIAN

The standard unit of *solid angular measure* is the *steradian*, symbolized sr. A solid angle of 1 sr is represented by a cone with its apex at the center of a sphere, intersecting the surface of the sphere in a circle such that, within the circle, the enclosed area on the sphere is equal to the square of the radius of the sphere (Fig. 4-7). There are 4π, or approximately 12.56636, steradians in a complete sphere.

THE NEWTON

The standard unit of *mechanical force* is the *newton*, symbolized N. One newton is the amount of force that it takes to make a mass of 1 kg accelerate at a rate of one meter per second squared ($1\,\text{m/s}^2$). Jet or rocket engine

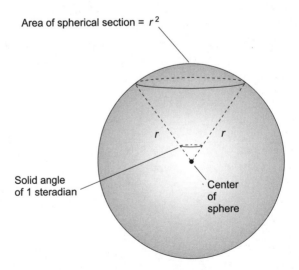

Fig. 4-7.   One steradian is a solid angle that defines an area on a sphere equal to the square of the radius of the sphere.

propulsion is measured in newtons. Force is equal to the product of mass and acceleration; reduced to base units in SI, newtons are equivalent to kilogram-meters per second squared $(kg \cdot m/s^2)$.

THE JOULE

The standard unit of *energy* is the *joule*, symbolized J. This is a small unit in real-world terms. One joule is the equivalent of a newton-meter $(N \cdot m)$. If reduced to base units in SI, the joule can be expressed in terms of unit mass multiplied by unit distance squared per unit time squared:

$$1\,J = 1 \ kg \cdot m^2/s^2$$

THE WATT

The standard unit of *power* is the *watt*, symbolized W. One watt is equivalent to one joule of energy expended per second of time (1 J/s). Power is a measure of the rate at which energy is produced, radiated, or consumed. The expression of watts in terms of SI base units begins to get esoteric:

$$1\,W = 1 \ kg \cdot m^2/s^3$$

You won't often hear about an electric bulb being rated at "60 kilogram-meters squared per second cubed," but technically, this is equivalent to saying that the bulb is rated at 60 W.

Various units smaller or larger than the watt are often employed to measure or define power. A *milliwatt* (mW) is 0.001 W. A *microwatt* (μW) is 0.000001 W or 10^{-6} W. A *nanowatt* (nW) is 10^{-9} W. A *kilowatt* (kW) is 1000 W. A *megawatt* (MW) is 1,000,000 W or 10^6 W.

THE COULOMB

The standard unit of *electric charge quantity* is the *coulomb*, symbolized C. This is the electric charge that exists in a congregation of approximately 6.241506×10^{18} electrons. It also happens to be the electric charge contained in that number of protons, anti-protons, or positrons (anti-electrons). When you walk along a carpet with hard-soled shoes in the winter, or anywhere the humidity is very low, your body builds up a static electric charge that can be expressed in coulombs (or more likely a fraction of one coulomb). Reduced to base units in SI, one coulomb is equal to one ampere-second $(1\,A \cdot s)$.

The coulomb, like the mole, is a dimensionless unit. It represents a numeric quantity. In fact, 1 C is equivalent to approximately 1.0365×10^{-5} mol. The coulomb is always used in reference to quantity of particles that carry a *unit electric charge*; but that is simply a matter of convention.

THE VOLT

The standard unit of *electric potential* or *potential difference*, also called *electromotive force* (EMF), is the *volt*, symbolized V. One volt is equivalent to one joule per coulomb (1 J/C). The volt is, in real-world terms, a moderately small unit of electric potential.

A standard dry cell of the sort you find in a flashlight (often erroneously called a "battery"), produces about 1.5 V. Most automotive batteries in the United States produce between 12 V and 13.5 V. Some high-powered radio and audio amplifiers operate with voltages in the hundreds or even thousands.

You'll sometimes encounter expressions of voltage involving prefix multipliers. A *millivolt* (mV) is 0.001 V. A *microvolt* (μV) is 0.000001 V or 10^{-6} V. A *nanovolt* (nV) is 10^{-9} V. A *kilovolt* (kV) is 1000 V. A *megavolt* (MV) is 1,000,000 V or 10^6 V.

THE OHM

The standard unit of *electrical resistance* is the *ohm*, symbolized by the uppercase Greek letter omega (Ω) or sometimes simply written out in full as "ohm." Resistance is mathematically related to the current and the voltage in any direct-current (dc) electrical circuit.

When a current of 1 A flows through a resistance of 1 Ω a voltage of 1 V appears across that resistance. If the current is doubled to 2 A and the resistance remains at 1 Ω, the voltage doubles to 2 V; if the current is cut by a factor of 4 to 0.25 A, the voltage likewise drops by a factor of 4 to 0.25 V. The ohm is thus equivalent to a volt per ampere (1 V/A).

The ohm is a small unit of resistance, and you will often see resistances expressed in units of thousands or millions of ohms. A *kilohm* (kΩ or k) is 1000 Ω; a *megohm* (MΩ or M) is 1,000,000 Ω.

THE HERTZ

The standard unit of *frequency* is the *hertz*, symbolized Hz. It was formerly called the *cycle per second* or simply the *cycle*. If a wave has a frequency of 1 Hz, it goes through one complete cycle every second. If the frequency is 2 Hz, the wave goes through two cycles every second, or one cycle every

1/2 second. If the frequency is 10 Hz, the wave goes through 10 cycles every second, or one cycle every 1/10 second.

The hertz is used to express audio frequencies, for example the pitch of a musical tone. The hertz is also used to define the frequencies of wireless signals, both transmitted and received. It can even represent signal bandwidth.

The hertz is a small unit in the real world, and 1 Hz represents an extremely low frequency. More often, you will hear about frequencies that are measured in thousands, millions, billions (thousand-millions), or trillions (million-millions) of hertz. These units are called *kilohertz* (kHz), *megahertz* (MHz), *gigahertz* (GHz), and *terahertz* (THz), respectively.

In terms of SI units, the hertz is mathematically simple, but the concept is esoteric for some people to grasp: it is an "inverse second" (s^{-1}) or "per second" (/s).

PROBLEM 4-4

It has been said that the clock speed of the microprocessor in an average personal computer doubles every year. Suppose that is precisely true, and continues to be the case for at least a decade to come. If the average microprocessor clock speed on January 1, 2005 is 10 GHz, what will be the average microprocessor clock speed on January 1, 2008?

SOLUTION 4-4

The speed will double 3 times; that means it will be 2^3, or 8, times as great. Thus, on January 1, 2008, the average personal computer will have a microprocessor rated at 10 GHz × 8 = 80 GHz.

PROBLEM 4-5

In the preceding scenario, what will be the average microprocessor clock speed on January 1, 2015?

SOLUTION 4-5

In this case the speed will double 10 times; it will become 2^{10}, or 1024, times as great. That means the speed will be 10 GHz × 1024, or 10,240 GHz. This is better expressed as 10.240 THz which, because we are told the original speed to only two significant figures, is best rounded to an even 10 THz.

Conversions

With all the different systems of units in use throughout the world, the business of conversion from one system to another has become the subject matter for Web sites. Nevertheless, in order to get familiar with how units relate to each other, it's a good idea to do a few manual calculations before going online and letting your computer take over the "dirty work." The problems below are some examples; you can certainly think of other unit-conversion situations that you are likely to encounter in your everyday affairs. Table 4-1 can serve as a guide for converting base units.

DIMENSIONS

When converting from one unit system to another, always be sure you're talking about the same quantity or phenomenon. For example, you cannot convert meters squared to centimeters cubed, or candela to meters per second. You must keep in mind what you're trying to express, and be sure you are not, in effect, trying to change an apple into a drinking glass.

The particular thing that a unit quantifies is called the *dimension* of the quantity or phenomenon. Thus, meters per second, feet per hour, and furlongs per fortnight represent expressions of the speed dimension; seconds, minutes, hours, and days are expressions of the time dimension.

PROBLEM 4-6
Suppose you step on a scale and it tells you that you weigh 120 pounds. How many kilograms does that represent?

SOLUTION 4-6
Assume you are on the planet earth, so your mass-to-weight conversion can be defined in a meaningful way. (Remember, mass is not the same thing as weight.) Use Table 4-1. Multiply by 0.4535 to get 54.42 kg. Because you are given your weight to only three significant figures, you should round this off to 54.4 kg.

PROBLEM 4-7
You are driving in Europe and you see that the posted speed limit is 90 kilometers per hour (km/hr). How many miles per hour (mi/hr) is this?

SOLUTION 4-7
In this case, you only need to worry about miles versus kilometers; the "per hour" part doesn't change. So you convert kilometers to miles. First remember that $1\,km = 1000\,m$; then $90\,km = 90,000\,m = 9.0 \times 10^4\,m$. The conversion of meters to statute miles (these are the miles used on land) requires that you multiply by 6.214×10^{-4}. Therefore, you multiply 9.0×10^4 by 6.214×10^{-4} to get 55.926. This must be rounded off to 56, or

Table 4-1 Conversions for base units in the International System (SI) to units in other systems. When no coefficient is given, it is exactly equal to 1.

To convert:	To:	Multiply by:	Conversely, multiply by:
meters (m)	nanometers (nm)	10^9	10^{-9}
meters (m)	microns (μ)	10^6	10^{-6}
meters (m)	millimeters (mm)	10^3	10^{-3}
meters (m)	centimeters (cm)	10^2	10^{-2}
meters (m)	inches (in)	39.37	0.02540
meters (m)	feet (ft)	3.281	0.3048
meters (m)	yards (yd)	1.094	0.9144
meters (m)	kilometers (km)	10^{-3}	10^3
meters (m)	statute miles (mi)	6.214×10^{-4}	1.609×10^3
meters (m)	nautical miles	5.397×10^{-4}	1.853×10^3
kilograms (kg)	nanograms (ng)	10^{12}	10^{-12}
kilograms (kg)	micrograms (μg)	10^9	10^{-9}
kilograms (kg)	milligrams (mg)	10^6	10^{-6}
kilograms (kg)	grams (g)	10^3	10^{-3}
kilograms (kg)	ounces (oz)	35.28	0.02834
kilograms (kg)	pounds (lb)	2.205	0.4535
kilograms (kg)	English tons	1.103×10^{-3}	907.0
seconds (s)	minutes (min)	0.01667	60.00
seconds (s)	hours (h)	2.778×10^{-4}	3.600×10^3
seconds (s)	days (dy)	1.157×10^{-5}	8.640×10^4

(Continued)

Table 4-1　Continued.

To convert:	To:	Multiply by:	Conversely, multiply by:
seconds (s)	years (yr)	3.169×10^{-8}	3.156×10^{7}
degrees Kelvin (°K)	degrees Celsius (°C)	Subtract 273	Add 273
degrees Kelvin (°K)	degrees Fahrenheit (°F)	Multiply by 1.80, then subtract 459	Multiply by 0.556, then add 255
degrees Kelvin (°K)	degrees Rankine (°R)	1.80	0.556
amperes (A)	carriers per second	6.24×10^{18}	1.60×10^{-19}
amperes (A)	nanoamperes (nA)	10^{9}	10^{-9}
amperes (A)	microamperes (µA)	10^{6}	10^{-6}
amperes (A)	milliamperes (mA)	10^{3}	10^{-3}
candela (cd)	microwatts per steradian (µW/sr)	1.464×10^{3}	6.831×10^{-4}
candela (cd)	milliwatts per steradian (mW/sr)	1.464	0.6831
candela (cd)	watts per steradian (W/sr)	1.464×10^{-3}	683.1
moles (mol)	coulombs (C)	9.65×10^{4}	1.04×10^{-5}

two significant figures, because the posted speed limit quantity, 90, only goes that far.

PROBLEM 4–8
How many feet per second is the above-mentioned speed limit? Use the information in Table 4-1.

SOLUTION 4-8
Let's convert kilometers per hour to kilometers per second first. This requires division by 3600, the number of seconds in an hour. Thus, 90 km/hr = 90/3600 km/s = 0.025 km/s. Next, convert kilometers to meters. Multiply by

1000 to obtain 25 m/s as the posted speed limit. Finally, convert meters to feet. Multiply 25 by 3.281 to get 82.025. This must be rounded off to 82 ft/sec because the posted speed limit is expressed to only two significant figures.

Quiz

Refer to the text in this chapter if necessary. A good score is eight correct. Answers are in the back of the book.

1. Which of the following terms commonly represents more than one measurable quantity or phenomenon?
    (a) meter
    (b) pound
    (c) radian
    (d) degree

2. The mole is a unit that expresses the
    (a) number of electrons in an ampere
    (b) number of particles in a sample
    (c) distance from the sun to a planet
    (d) time required for an electron to orbit an atomic nucleus

3. The nomograph of Fig. 4-8 can be used to convert:
    (a) feet to meters
    (b) inches to centimeters
    (c) miles to kilometers
    (d) grams to pounds

4. A joule is the equivalent of a
    (a) foot-pound
    (b) meter per second
    (c) kilogram per meter
    (d) watt-second

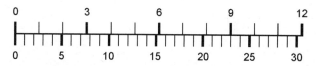

Fig. 4-8.   Illustration for Quiz Question 3.

5. An example of a dimensionless unit is
    (a) the kilogram
    (b) the meter
    (c) the second
    (d) none of the above

6. Force is equivalent to
    (a) mass multiplied by distance
    (b) mass multiplied by time
    (c) mass multiplied by acceleration
    (d) mass multiplied by speed

7. Two islands are 1345 nautical miles apart. How many kilometers is
    this?
    (a) 2492
    (b) 2164
    (c) 1853
    (d) It is impossible to say without more information.

8. The number of kilograms an object has can be converted to pounds
    only if you also know the
    (a) temperature of the object
    (b) size of the object
    (c) gravitational strength on the planet where the object is located
    (d) number of moles of atoms in the object

9. The pound is a unit of
    (a) mass
    (b) distance
    (c) acceleration
    (d) none of the above

10. The SI system is an expanded form of the
    (a) English system
    (b) metric system
    (c) European system
    (d) American system

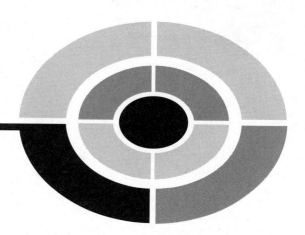

Test: Part 1

Do not refer to the text when taking this test. You may draw diagrams or use a calculator if necessary. A good score is at least 30 correct. Answers are in the back of the book. It's best to have a friend check your score the first time, so you won't memorize the answers if you want to take the test again.

1.   The binary number 10010 is equivalent to the decimal number
     (a) 8
     (b) 12
     (c) 16
     (d) 18
     (e) 34

2.   In the graph of Fig. Test 1-1, how many orders of magnitude are shown on the horizontal axis?
     (a) 1
     (b) 2
     (c) 10
     (d) 100
     (e) Infinitely many.

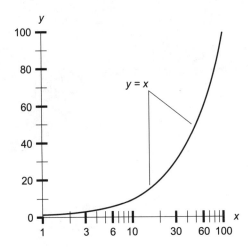

Fig. Test 1-1. Illustration for Part One Test Questions 2 and 3.

3. In Fig. Test 1-1, why does the graph of $y = x$ look like a curve, not a straight line?
   (a) Because one scale is linear (graduated evenly) and the other is not.
   (b) Because neither scale is linear.
   (c) Because both scales are linear.
   (d) Because y is not a function of x in this example.
   (e) Because the two scales encompass the same range of values.

4. Suppose you graph the barometric pressure at 30-minute intervals during the passage of a hurricane, with time on the horizontal scale and pressure (in millibars) on the vertical scale. After you have plotted the points, you connect them with a curve that represents your "best educated guess" as to the moment-to-moment pressure. In this graph, the pressure is
   (a) the independent variable
   (b) the dependent variable
   (c) a function
   (d) a relation
   (e) a coefficient

5. In the graph described in the previous question, suppose you connect the points with straight lines to get an approximate graph of the barometric pressure plotted with respect to time. This lets you infer the pressure at any given point in time according to
   (a) extrapolation
   (b) inversion

(c) linear interpolation

(d) exponentiation

(e) curve fitting

6. In the graph described in Question 4, suppose you connect the points with a "best educated guess" curve in an attempt to create a precise, moment-to-moment graph of the barometric pressure plotted with respect to time. This lets you infer the pressure at any given point in time according to
   (a) extrapolation
   (b) inversion
   (c) linear interpolation
   (d) exponentiation
   (e) curve fitting

7. Consider a number that is equal to 5.00×2^{20}. In power-of-10 notation, rounded to three significant figures, this is
   (a) 5.24×10^6
   (b) 2.00×10^3
   (c) 5.12×10^3
   (d) 1.00×10^{20}
   (e) impossible to determine without more information

8. Fig. Test 1-2 shows the speed of an accelerating car (in meters per second) as a function of time (in seconds). The fact that the graph is a straight line indicates that the rate of acceleration (expressed in meters per second per second, or m/s^2) is constant. What is this constant acceleration?
   (a) 1 m/s^2
   (b) 2 m/s^2
   (c) 10 m/s^2
   (d) 20 m/s^2
   (e) It cannot be determined without more information.

9. In the scenario shown by Fig. Test 1-2, suppose a mass of one kilogram (1 kg) is substituted for the car. According to the laws of physics, constant acceleration of an object having constant mass is produced by a constant force or thrust. Force is expressed in units of kilogram-meters per second per second ($kg \cdot m/s^2$), also known as newtons (N), as you learned in Chapter 4. What force, in newtons, is necessary to cause a 1-kg mass to accelerate at the rate shown by the graph?
   (a) 1 N
   (b) 2 N

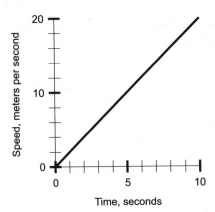

Fig. Test 1-2.   Illustration for Part One Test Questions 8 and 9.

    (c) 10 N

    (d) 20 N

    (e) It cannot be determined without more information.

10.   In a graph intended to show the correlation between two variables, the total absence of any correlation is indicated by points that

    (a) all lie near a horizontal line

    (b) all lie near a vertical line

    (c) are scattered all over the graph

    (d) all lie tightly clustered near the center of the graph

    (e) are arranged in some manner other than the four described above

11.   How does a numeral differ from a number?

    (a) A number represents a diagram, but a numeral represents a set.

    (b) A numeral is a symbol that represents a number.

    (c) A number is a symbol that represents a numeral.

    (d) Numerals are letters of the alphabet, but numbers are symbols such as the plus sign, the minus sign, and the forward slash.

    (e) There is no difference.

12.   What is the value of $3 - 5 \times 6/3 + 10$?

    (a) 6

    (b) $-9/10$

    (c) -40

    (d) 3

    (e) It is undefined, because it is ambiguous as written.

13.   In a vertical bar graph, function values are portrayed as

    (a) the heights of bars having equal width

(b) the widths of bars having equal height

(c) points connected by straight lines

(d) points connected by a curve

(e) points plotted all by themselves, not connected by anything

14. The milligram is a unit of
    (a) temperature
    (b) electrical quantity
    (c) brightness of light
    (d) frequency
    (e) mass

15. The mole is an example of a
    (a) linear unit
    (b) nonlinear unit
    (c) dimensional unit
    (d) dimensionless unit
    (e) temporal unit

16. The sum of two integers
    (a) is always an integer
    (b) is always a positive integer
    (c) cannot be a rational number
    (d) is never zero
    (e) is always a dependent variable

17. A radian is equivalent to approximately
    (a) 57.3 meters of displacement
    (b) 57.3 degrees of angular measure
    (c) 57.3 degrees Kelvin
    (d) 57.3 degrees Fahrenheit
    (e) 57.3 degrees Celsius

18. In order to get the Fahrenheit (°F) temperature when the Kelvin temperature (°K) is known, this formula is commonly used:

$$°F = 1.800 \times °K - 459.67$$

Suppose we are told that the temperature outdoors is 300 °K. What is the temperature in °F to the maximum justifiable number of significant figures?
    (a) 80.3 °F
    (b) 80.33 °F
    (c) −287 °F

(d) −287.4 °F

(e) None of the above.

19. If the independent and dependent variables of a function are reversed, the result

(a) is always a function

(b) is sometimes a function

(c) is never a function

(d) is not defined

(e) has negative correlation

20. Suppose there are two quantities, and one of them is exactly 0.001 the size of the other. These quantities differ by

(a) 0.001 orders of magnitude

(b) 1 order of magnitude

(c) 3 orders of magnitude

(d) 10 orders of magnitude

(e) 1000 orders of magnitude

21. In the octal system, the number 67 is followed by the number

(a) 68

(b) 69

(c) 70

(d) 77

(e) 100

22. The numbers $-6.78870000 \times 10^{-6}$ and -6.7887×10^{-6} differ in the sense that

(a) the first is expressed with a greater degree of accuracy than the second

(b) the first is four orders of magnitude larger than the second

(c) the first is only a small fraction of the second

(d) the first is rounded, but the second is truncated

(e) the first is the reciprocal of the second

23. Every rational number is

(a) a natural number

(b) an integer

(c) a real number

(d) positive

(e) nonzero

24. Consider that a meter (m) is equivalent to 39.37 inches (in), a foot (ft) is equivalent to 12.000 in, and a statute mile (mi) is equivalent

to 5280 ft. Based on this information, how many feet are there in a kilometer?
(a) 3281 ft
(b) 1609 ft
(c) 440.0 ft
(d) 1.732×10^4 ft
(e) More information is needed to calculate this.

25. Based on the information given in the previous question, how many millimeters (mm) are there in a statute mile?
(a) 6.336×10^4 mm
(b) 1.341×10^5 mm
(c) 1.609×10^6 mm
(d) 6.336×10^7 mm
(e) More information is needed to calculate this.

26. In the hexadecimal system, the number 11 is followed by the number
(a) 12
(b) A
(c) B
(d) 1A
(e) 1B

27. How many radians are there in a degree Kelvin?
(a) Approximately 57.3, the same number of degrees as there are in a radian.
(b) Approximately 273.15, the difference between °K and °C.
(c) Approximately 360, the number of degrees in a complete circle.
(d) It depends on the radius of the circle under consideration.
(e) This question is meaningless because the units are incompatible.

28. Suppose a and b are rational numbers, both larger than 1 but less than 10, and both expressed in decimal form to the same number of significant figures. Suppose m and n are integers. What is the product of $a \times 10^m$ and $b \times 10^n$?
(a) There is no way to express this in the general case.
(b) $(a+b) \times 10^{mn}$
(c) $ab \times 10^{mn}$
(d) $ab \times (10m)^n$
(e) $ab \times 10^{(m+n)}$

29. The steradian is a unit of
(a) solid angular measure

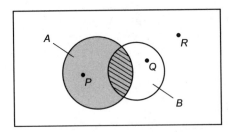

Fig. Test 1-3. Illustration for Part One Test Questions 30 and 31.

    (b) temperature
    (c) visible-light brightness
    (d) frequency
    (e) electrical current

30.   In Fig. Test 1-3, the gray shaded area represents
    (a) the set $A - B$
    (b) the set $A \cap B$
    (c) the set $A \cup B$
    (d) the set $A \subseteq B$
    (e) the set A

31.   In Fig. Test 1-3, which of the points P, Q, or R lies within the set $A \cap B$?
    (a) Point P only.
    (b) Point Q only.
    (c) Points P and R.
    (d) Points Q and R.
    (e) None of the points.

32.   In Fig. Test 1-4, the point shown represents
    (a) a natural number
    (b) an integer
    (c) a rational number
    (d) an irrational number
    (e) none of the above

33.   A one-dimensional graph, consisting of two scales lined directly up against each other, is called a
    (a) linear comparison
    (b) bar graph
    (c) correlation graph
    (d) nomograph
    (e) functional graph

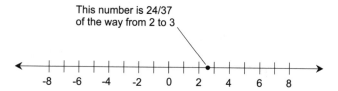

Fig. Test 1-4. Illustration for Part One Test Question 32.

34. One millionth of a kilometer is equal to
    (a) 1 meter
    (b) 10 meters
    (c) 1 millimeter
    (d) 10 millimeters
    (e) 1 micrometer

35. In a graph intended to show the correlation between two variables, strong negative correlation is indicated by points that
    (a) all lie near a horizontal line
    (b) all lie near a vertical line
    (c) are scattered all over the graph
    (d) all lie tightly clustered near the center of the graph
    (e) are arranged in some manner other than the four described above

36. Suppose you are performing calculations on a computer, and you end up with the expression 5.000000000E−45. This is
    (a) a number between 0 and 1
    (b) a number greater than 1
    (c) a number between −1 and 0
    (d) a number less than −1
    (e) a number too large to be defined

37. The reciprocal of 5.000000000E−45 can be expressed as
    (a) 2×10^{-44}
    (b) 2×10^{44}
    (c) 5×10^{45}
    (d) 5×10^{44}
    (e) none of the above

38. In Fig. Test 1-5, the solid curve represents an approximation of stock price versus time, based on points plotted
    (a) continuously
    (b) at regular time intervals

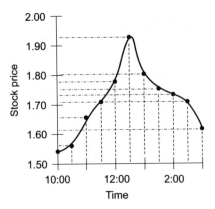

Fig. Test 1-5. Illustration for Part One Test Questions 38 and 39.

    (c) at regular stock-price intervals
    (d) at random
    (e) by extrapolation

39.   In Fig. Test 1-5, suppose stock price is considered as a function of time. We can call this
    (a) a nondecreasing function
    (b) a nonincreasing function
    (c) both a nondecreasing and a nonincreasing function
    (d) neither a nondecreasing nor a nondecreasing function
    (e) a relation, but not a legitimate function

40.   In the hexadecimal number system, the single-digit numerals are
    (a) 1 and 2
    (b) 1, 2, 3, 4, 5, 6, 7, 8, 9, and 10
    (c) 1, 2, 3, 4, 5, 6, 7, and 8
    (d) 1, 2, 3, and 4
    (e) none of the above

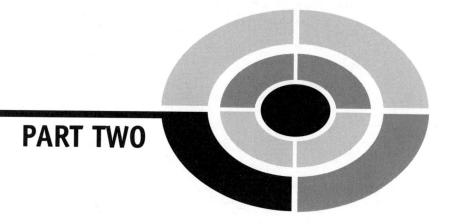

PART TWO

Finding Unknowns

Basic Algebra

Linear equations have variables that are not raised to any power (except 1 or 0). They are the simplest equations to solve, although when there are four or more linear equations in four or more variables, the solution process can be messy. In this chapter, we'll deal with linear equations in one and two variables.

Single-Variable Linear Equations

The objective of solving a single-variable equation is to get it into a form where the expression on the left-hand side of the equality symbol is exactly equal to the variable being sought (for example, x), and a defined expression *not containing that variable* is on the right.

ELEMENTARY RULES

There are several ways in which an equation in one variable can be manipulated to obtain a solution, assuming a solution exists. Any and all of the principles outlined in Chapter 1 can be applied toward this result. In addition, the following rules can be applied in any order, and any number of times.

Addition of a quantity to each side: Any defined constant, variable, or expression can be added to both sides of an equation, and the result is equivalent to the original equation.

Subtraction of a quantity from each side: Any defined constant, variable, or expression can be subtracted from both sides of an equation, and the result is equivalent to the original equation.

Multiplication of each side by a nonzero quantity: Both sides of an equation can be multiplied by a nonzero constant, variable, or expression, and the result is equivalent to the original equation.

Division of each side by a quantity: Both sides of an equation can be divided by a nonzero constant, by a variable that cannot attain a value of zero, or by an expression that cannot attain a value of zero over the range of its variable(s), and the result is equivalent to the original equation.

BASIC EQUATION IN ONE VARIABLE

Consider an equation of the following form:

$$ax + b = cx + d$$

where a, b, c, and d are real numbers, and $a \neq c$. This equation is solved as follows:

$$ax + b = cx + d$$
$$ax = cx + d - b$$
$$ax - cx = d - b$$
$$(a - c)x = d - b$$
$$x = (d - b)/(a - c)$$

STANDARD FORM

Any single-variable linear equation can be reduced to this form, called the *standard form*:

$$ax + b = 0$$

where a and b are constants, and x is the variable. The value on the right-hand side of the equals sign is always 0 when a linear equation is in standard form.

ILLUSTRATING THE SOLUTION

Suppose we substitute y for 0 in the standard form of a linear equation. This produces a relation where y is the dependent variable and x is the independent variable:

$$ax + b = y$$

Let's create a graph of this relation for particular values of a and b, say $a = 2$ and $b = 3$, so the linear equation is

$$2x + 3 = y$$

Put the x-values on the horizontal axis and the y-values on the vertical axis. Plot several test points by setting x at certain values, calculating the y values that result, and then plotting the points on the graph. You should get something that looks like Fig. 5-1. The graph is a straight line. Try this for several different combinations of constants a and b. You will find that the graph always turns out as a straight line. That's where the term "linear equation" comes from!

The solution to the single-variable equation $ax + b = 0$ is indicated by the point where the straight-line graph for $ax + b = y$ crosses the x axis (that is,

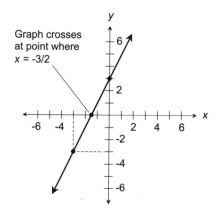

Fig. 5-1.   When the equation $2x + 3 = y$ is plotted as a graph, it turns out to be a straight line.

the horizontal axis). That is, it is the value of x for the particular case where $y=0$. In Fig. 5-1, this happens to be $x=-3/2$.

PROBLEM 5-1
Solve the following equation for x. Show the results on a graph similar to Fig. 5-1, by replacing the 0 on the right-hand side of the equation by the variable y.

$$4x + 3 = 0$$

SOLUTION 5-1
To solve $4x+3=0$, proceed like this:

$$4x + 3 = 0$$
$$4x = -3 \text{ (subtract 3 from each side)}$$
$$x = -3/4 \text{ (divide each side by 4)}$$

The graph of the equation $4x+3=y$ is shown in Fig. 5-2. Note where the straight line crosses the x axis.

PROBLEM 5-2
Solve the following equation for x. Show the results on a graph similar to Fig. 5-1, by reducing the equation to the standard form $ax+b=0$, and then replacing the 0 on the right-hand side of the equation by the variable y.

$$7x - 4 = 3x + 1$$

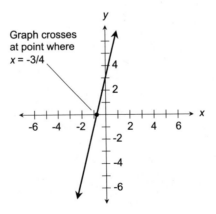

Fig. 5-2.   Illustration for Problem 5-1.

SOLUTION 5-2

To solve $7x - 4 = 3x + 1$, proceed like this:

$$7x - 4 = 3x + 1$$

$$7x = 3x + 5 \text{ (add 4 to each side)}$$

$$4x = 5 \text{ (subtract } 3x \text{ from each side)}$$

$$x = 5/4 \text{ (divide each side by 4)}$$

To reduce the equation to the standard form, proceed this way:

$$7x - 4 = 3x + 1$$

$$4x - 4 = 1 \text{ (subtract } 3x \text{ from each side)}$$

$$4x - 5 = 0 \text{ (subtract 1 from each side)}$$

The graph of the equation $4x - 5 = y$ is shown in Fig. 5-3. Note where the straight line crosses the x axis; this shows the solution $x = 5/4$.

PROBLEM 5-3

Solve the following equation for x. Show the results on a graph similar to Fig. 5-1, by replacing the 0 on the right-hand side of the equation by the variable y.

$$-2x - 4 = 3x$$

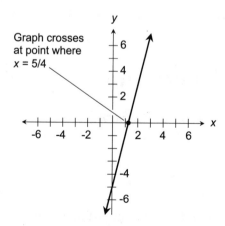

Fig. 5-3.   Illustration for Problem 5-2.

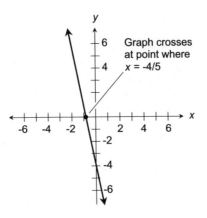

Fig. 5-4.   Illustration for Problem 5-3.

SOLUTION 5-3

To solve $-2x - 4 = 3x$, proceed like this:

$$-2x - 4 = 3x$$
$$-4 = 5x \text{ (add } 2x \text{ to each side)}$$
$$-4/5 = x \text{ (divide each side by 6)}$$
$$x = -4/5 \text{ (switch sides)}$$

To reduce the equation to the standard form, perform this single step:

$$-2x - 4 = 3x$$
$$-5x - 4 = 0 \text{ (subtract } 3x \text{ from each side)}$$

The graph of the equation $-5x - 4 = y$ is shown in Fig. 5-4. Note where the straight line crosses the x axis; this shows the solution $x = -4/5$.

Two-by-Two Linear Equations

A pair of linear equations in two variables, called a *two-by-two* (2×2) *set of linear equations*, takes the following general form:

$$a_1 x + b_1 y + c_1 = 0$$
$$a_2 x + b_2 y + c_2 = 0$$

where x and y represent the variables, and the a's, b's, and c's represent constants.

2 × 2 SUBSTITUTION METHOD

Consider a 2×2 set of linear equations in the above form. The substitution method of solving these equations consists in performing either of the following sequences of steps. If $a_1 \neq 0$, use Sequence A. If $a_1 = 0$, use Sequence B. If both $a_1 = 0$ and $a_2 = 0$, the set of equations is in fact a pair of equations in terms of a single variable, and the following steps are irrelevant.

Sequence A. First, solve the first equation for x in terms of y:

$$a_1 x + b_1 y + c_1 = 0$$
$$a_1 x = -b_1 y - c_1$$
$$x = (-b_1 y - c_1)/a_1$$

Next, substitute the above-derived solution for x in place of x in the second equation from the original pair, obtaining:

$$a_2[(-b_1 y - c_1)/a_1] + b_2 y + c_2 = 0$$

Solve this single-variable equation for y, using the previously outlined rules for solving single-variable equations. Assuming a solution exists, it can be substituted for y in either of the original equations, deriving a single-variable equation in terms of x. Solve for x, using the previously outlined rules for solving single-variable equations.

Sequence B. Because $a_1 = 0$, the first equation has only one variable, and is in the following form:

$$b_1 y + c_1 = 0$$

Solve this equation for y:

$$b_1 y = -c_1$$
$$y = -c_1/b_1$$

This can be substituted for y in the second equation from the original pair, obtaining:

$$a_2 x + b_2(-c_1/b_1) + c_2 = 0$$
$$a_2 x - b_2(c_1/b_1) + c_2 = 0$$
$$a_2 x = b_2(c_1/b_1) - c_2$$
$$x = [b_2(c_1/b_1) - c_2]/a_2$$

2 × 2 ADDITION METHOD

Consider again the following set of two linear equations in two variables:

$$a_1x + b_1y + c_1 = 0$$
$$a_2x + b_2y + c_2 = 0$$

where a_1, a_2, b_1, b_2, c_1, and c_2 are constants, and the variables are represented by x and y. The addition method involves performing two separate and independent steps:

- Multiply one or both equations through by constant values to cancel out the coefficients of x, and then solve for y.
- Multiply one or both equations through by constant values to cancel out the coefficients of y, and then solve for x.

The scheme for solving for y begins by multiplying the first equation through by $-a_2$, and the second equation through by a_1, and then adding the two resulting equations:

$$-a_2a_1x - a_2b_1y - a_2c_1 = 0$$
$$a_1a_2x + a_1b_2y + a_1c_2 = 0$$
$$\overline{}$$
$$(a_1b_2 - a_2b_1)y + a_1c_2 - a_2c_1 = 0$$

Next, add a_2c_1 to each side, obtaining:

$$(a_1b_2 - a_2b_1)y + a_1c_2 = a_2c_1$$

Next, subtract a_1c_2 from each side, obtaining:

$$(a_1b_2 - a_2b_1)y = a_2c_1 - a_1c_2$$

Finally, divide through by $a_1b_2 - a_2b_1$, obtaining:

$$y = (a_2c_1 - a_1c_2)/(a_1b_2 - a_2b_1)$$

For this to be valid, the denominator must be nonzero. This means the products a_1b_2 and a_2b_1 must not be equal to each other.

The process of solving for x is similar. Consider again the original set of simultaneous linear equations:

$$a_1x + b_1y + c_1 = 0$$
$$a_2x + b_2y + c_2 = 0$$

Multiply the first equation through by $-b_2$, and the second equation through by b_1, and then add the two resulting equations:

$$-a_1b_2x - b_1b_2y - b_2c_1 = 0$$
$$a_2b_1x + b_1b_2y + b_1c_2 = 0$$
$$\overline{(a_2b_1 - a_1b_2)x + b_1c_2 - b_2c_1 = 0}$$

Next, add b_2c_1 to each side, obtaining:

$$(a_2b_1 - a_1b_2)x + b_1c_2 = b_2c_1$$

Next, subtract b_1c_2 from each side, obtaining:

$$(a_2b_1 - a_1b_2)x = b_2c_1 - b_1c_2$$

Finally, divide through by $a_2b_1 - a_1b_2$, obtaining:

$$x = (b_2c_1 - b_1c_2)/(a_2b_1 - a_1b_2)$$

For this to be valid, the denominator must be nonzero; this means $a_1b_2 \neq a_2b_1$.

WHERE TWO LINES CROSS

When you have a pair of linear equations in two variables, you can graph them both, and they will always both appear as straight lines. Try this with several different pairs of linear equations, letting the horizontal axis show x values and the vertical axis show y values. Usually, the two lines intersect at some point, although once in a while they are parallel, and don't intersect anywhere.

The solution to a pair of linear equations in two variables (if a solution exists) corresponds to the x and y values of the intersection point for the lines representing the equations. If the two lines are parallel, then there is no solution.

Graphing a pair of linear equations is not a good way to find the solution, or even of determining whether or not a solution exists. This is because graphing is a tedious and approximate business. It's a lot of work to generate the graphs for a pair of equations, and once you have drawn them, figuring out where (or even if) the lines intersect is subject to estimation errors. It's easiest to solve 2×2 sets of linear equations using the addition method. That's the scheme that we'll use in the next two problems. We'll graph them, too, but only as an aid to visualizing the situations.

PROBLEM 5-4

Use the 2 × 2 addition method to solve the following pair of linear equations, reducing them to standard form and then "plugging in" the constants to the general equations for x and y derived above.

$$2x - y = 2$$
$$-x + 3y = -3$$

SOLUTION 5-4

To convert the first equation to standard form, subtract 2 from each side. To convert the second equation to standard form, add 3 to each side. This gives us the following:

$$2x - y - 2 = 0$$
$$-x + 3y + 3 = 0$$

The constants in the first equation are $a_1 = 2$, $b_1 = -1$, and $c_1 = -2$. The constants in the second equation are $a_2 = -1$, $b_2 = 3$, and $c_2 = 3$. Therefore, we can solve for x by "plugging in" the appropriate constants to the following general equation:

$$
\begin{aligned}
x &= (b_2 c_1 - b_1 c_2)/(a_2 b_1 - a_1 b_2) \\
&= [3 \times (-2) - (-1) \times 3]/[(-1) \times (-1) - 2 \times 3] \\
&= [-6 - (-3)]/(1 - 6) \\
&= (-3)/(-5) \\
&= 3/5
\end{aligned}
$$

We can solve for y by "plugging in" the appropriate constants as follows:

$$
\begin{aligned}
y &= (a_2 c_1 - a_1 c_2)/(a_1 b_2 - a_2 b_1) \\
&= [(-1) \times (-2) - 2 \times 3]/[2 \times 3 - (-1) \times (-1)] \\
&= (2 - 6)/(6 - 1) \\
&= -4/5
\end{aligned}
$$

Thus, $x = 3/5$ and $y = -4/5$. Sometimes this is expressed by writing the numbers one after the other, between parentheses, separated by a comma, but without any space after the comma:

$$(x, y) = (3/5, -4/5)$$

PROBLEM 5-5
Use the 2×2 addition method to solve the original pair of equations from the previous problem directly, rather than by substituting constants into the general forms of the equations.

SOLUTION 5-5
Here are the original two equations again:

$$2x - y = 2$$
$$-x + 3y = -3$$

To solve for x, we must get the y terms to cancel out when we add the two equations. This can be accomplished by multiplying both sides of the top equation by 3 and then adding the two resulting equations together:

$$6x - 3y = 6$$
$$\underline{-x + 3y = -3}$$
$$5x = 3$$

Then divide both sides of this equation by 5 to get $x = 3/5$.
To solve for y, we must get the x terms to cancel out when we add the two equations. In order to make this happen, multiply both sides of the bottom equation by 2, and then add the two resulting equations together:

$$2x - y = 2$$
$$\underline{-2x + 6y = -6}$$
$$5y = -4$$

Then divide both sides by 5 to get $y = -4/5$.

PROBLEM 5-6
Draw the straight-line graphs from the two equations in Problems 5-4 and 5-5, showing their intersection point at $(x,y) = (3/5, -4/5)$.

SOLUTION 5-6
We haven't learned the quick-and-easy way to graph linear equations yet. That stuff is reserved for a later chapter. For now, just pick some numbers for x and plot the resulting points for y. Two points should do for both equations, because we know they'll be straight lines; from basic geometry we know that any straight line can be uniquely determined by two points. The result should look like Fig. 5-5.

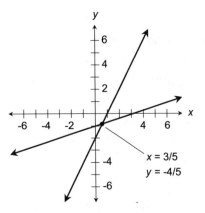

Fig. 5-5.   Illustration for Problem 5-6.

Quiz

Refer to the text in this chapter if necessary. A good score is eight correct. Answers are in the back of the book.

1. Which of the following is not a legitimate way to manipulate an equation in which the variable is *x*?
    (a) Subtract 1 from both sides.
    (b) Divide both sides by 0.
    (c) Multiply both sides by a constant.
    (d) Add 2*x* to both sides.

2. The term "linear," when used to describe an equation, comes from the fact that
    (a) such equations look like straight lines when graphed in two variables
    (b) the variables change at constant rates
    (c) such equations are straightforward and simple
    (d) the variables are actually constants

3. The solution to the equation $3x + 7 = -2x + 3$ is
    (a) undefined
    (b) equal to any and all real numbers
    (c) equal to $-4/5$
    (d) equal to $5/4$

4. Suppose the two straight lines in Fig. 5-6 represent linear equations in standard form, except that the 0 on the right-hand side of the

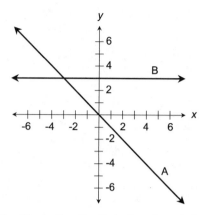

Fig. 5-6.   Illustration for Quiz Questions 4 through 7.

equals sign is replaced by y. Examine line A. It appears from this graph that when $y = 0$, the value of x

(a) is equal to 0

(b) is positive (greater than 0)

(c) is negative (less than 0)

(d) is not defined

5. Suppose the two straight lines in Fig. 5-6 represent linear equations in standard form, except that the 0 on the right-hand side of the equals sign is replaced by y. Examine line A. It appears from this graph that when y is positive (greater than 0), the value of x

(a) is equal to 0

(b) is positive (greater than 0)

(c) is negative (less than 0)

(d) is not defined

6. Suppose the two straight lines in Fig. 5-6 represent linear equations in standard form, except that the 0 on the right-hand side of the equals sign is replaced by y. Examine line B. It appears from this graph that when $y = 0$, the value of x

(a) can be any real number

(b) is positive (greater than 0)

(c) is equal to 0

(d) is not defined

7. Suppose the two straight lines in Fig. 5-6 represent linear equations in standard form, except that the 0 on the right-hand side of the equals

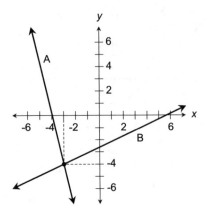

Fig. 5-7.   Illustration for Quiz Questions 8 and 9.

sign is replaced by y. Examine line B. It appears from this graph that when $y = 3$, the value of x
(a) can be any real number
(b) is positive (greater than 0)
(c) is equal to 0
(d) is not defined

8.   Examine Fig. 5-7. Suppose lines A and B represent a pair of linear equations in two variables, x and y. How many solutions exist to this pair of equations?
(a) None.
(b) One.
(c) Two.
(d) Infinitely many.

9.   Which of the following statements is true of any or all solutions (x, y) to the pair of linear equations shown by lines A and B in Fig. 5-7?
(a) $x = -3$
(b) $y = -3$
(c) $(x, y) = -3$
(d) $x + y = -3$

10.   What is the unique solution (x, y) to the equation $2x + 2y = 0$?
(a) $x = -2$
(b) $y = -2$
(c) $x + y = 0$
(d) An additional equation in x and y is necessary to answer this.

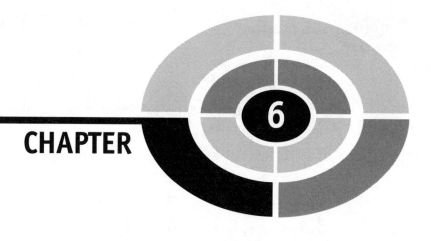

CHAPTER 6

More Algebra

When the variable, or unknown, in an equation is raised to a power, things get interesting. The larger the exponents attached to the variables, the more interesting things become, and the more difficult it gets to solve the equation. In this chapter, we'll look into the basics of quadratic, cubic, and higher-order equations in one variable. Even if you're not a mathematician or a scientist, you'll encounter equations like these once in a while, so it's good to know about them.

Quadratic Equations

A *one-variable, second-order equation*, also called a *second-order equation in one variable* or, more often, a *quadratic equation*, can be written in the following *standard form*:

$$ax^2 + bx + c = 0$$

where a, b, and c are constants, a is not equal to 0, and x is the variable. The constants a, b, and c are also known as *coefficients*.

SOME EXAMPLES

Any equation that can be converted into the above form is a quadratic equation. Alternative forms are:

$$mx^2 + nx = p$$

$$qx^2 = rx + s$$

$$(x + t)(x + u) = 0$$

where m, n, p, q, r, s, t, and u are constants. Here are some examples of quadratic equations in various forms:

$$x^2 + 2x + 1 = 0$$

$$-3x^2 - 4x = 2$$

$$4x^2 = -3x + 5$$

$$(x + 4)(x - 5) = 0$$

GET IT INTO FORM

Some quadratic equations are easy to solve. Others are difficult. The first step in finding the value(s) of the variable in a quadratic is to get the equation either into standard form or into *factored form.*

The first equation above is already in standard form. It is ready for an attempt at solution, which, as we will shortly see, is easy.

The second equation above can be reduced to standard form by subtracting 2 from each side:

$$-3x^2 - 4x = 2$$

$$-3x^2 - 4x - 2 = 0$$

The third equation above can be reduced to standard form by adding $3x$ to each side and then subtracting 5 from each side:

$$4x^2 = -3x + 5$$

$$4x^2 + 3x = 5$$

$$4x^2 + 3x - 5 = 0$$

An interesting aside: these are not the only ways the equations can be reduced to standard form. They can be multiplied through by any nonzero constant, and the resulting equations are still in standard form.

The fourth equation above is in factored form. This is a convenient form, because a quadratic equation denoted this way can be solved without having to do any work. Look at it closely:

$$(x + 4)(x - 5) = 0$$

The expression on the left-hand side of the equals sign is zero if either of the two factors is zero. If $x = -4$, then here is what happens when we "plug it in":

$$(-4 + 4)(-4 - 5) = 0$$
$$0 \times -9 = 0 \text{ (It works!)}$$

If $x = 5$, then here's what occurs:

$$(5 + 4)(5 - 5) = 0$$
$$9 \times 0 = 0 \text{ (It works again!)}$$

It is obvious which values for the variable in a factored quadratic will work as solutions. Simply take the additive inverses (negatives) of the constants in each factor. It's so easy, in fact, that you must think there's a catch. There is, of course. Most quadratic equations are difficult to get neatly into factored form.

A COUPLE OF NITS

Here are a couple of little things that should be cleared up right away, so you never get confused about them.

Suppose you run across a quadratic like this:

$$x(x + 3) = 0$$

You might want to imagine this equation written out like this:

$$(x + 0)(x + 3) = 0$$

This makes it easy to see that the solutions are $x = 0$ or $x = -3$. Of course, adding 0 to a variable doesn't change anything except to make an equation longer. But it can clarify the process of solving some equations.

In case you forgot, at the beginning of this section it was mentioned that a quadratic equation might have only one real-number solution. Here is an example of the factored form of such an equation:

$$(x - 11)(x - 11) = 0$$

In standard form, the equation looks like this:

$$x^2 - 22x + 121 = 0$$

Some purists might say, "This equation has two real-number solutions, and they are both equal to 11."

PROBLEM 6-1

Convert the following quadratic equations into factored form with real-number coefficients:

$$x^2 - 2x - 15 = 0$$
$$x^2 + 4 = 0$$

SOLUTION 6-1

The first equation turns out to have a "clean" factored equivalent:

$$(x + 3)(x - 5) = 0$$

The second equation does not have any real-numbered solutions, so the problem, as stated, can't be solved. Getting the second equation into factored form requires that you know something about *complex numbers*, and we'll get into that subject in a moment. You can tell that something is peculiar about this equation if you subtract 4 from each side:

$$x^2 + 4 = 0$$
$$x^2 = -4$$

No real number can be substituted for x in this equation in order to make it a true statement. That is because no real number, when squared, produces a negative real-number result.

PROBLEM 6-2

Put the following factored equations into standard quadratic form:

$$(x + 5)(x - 1) = 0$$
$$x(x + 4) = 0$$

SOLUTION 6-2

Both of these can be converted to standard form by multiplying the factors. In the first case, it can be done in two steps:

$$(x + 5)(x - 1) = 0$$
$$x^2 - x + 5x + [5 \times (-1)] = 0$$
$$x^2 + 4x - 5 = 0$$

In the second case, it can be done in a single step:

$$x(x + 4) = 0$$
$$x^2 + 4x = 0$$

THE QUADRATIC FORMULA

Examine these two quadratic equations:

$$-3x^2 - 4x = 2$$
$$4x^2 = -3x + 5$$

These can be reduced to standard form:

$$-3x^2 - 4x - 2 = 0$$
$$4x^2 + 3x - 5 = 0$$

You might stare at these equations for a long time before you get any ideas about how to factor them. Eventually, you might wonder why you are wasting your time. These equations do not "want" to be factored. Fortunately, there is a formula you can use to solve quadratic equations in general.

Consider the general quadratic equation:

$$ax^2 + bx + c = 0$$

where $a \neq 0$. The solution(s) to this equation can be found using this formula:

$$x = [-b \pm (b^2 - 4ac)^{1/2}]/2a$$

The symbol $\pm$ is read "plus-or-minus" and is a way of compacting two mathematical expressions into one. Written separately, the equations are:

$$x = [-b + (b^2 - 4ac)^{1/2}]/2a$$
$$x = [-b - (b^2 - 4ac)^{1/2}]/2a$$

The fractional exponent means the 1/2 power, another way of expressing the square root.

Plugging in

Examine this equation once again:

$$-3x^2 - 4x - 2 = 0$$

The coefficients are as follows:

$$a = -3$$
$$b = -4$$
$$c = -2$$

Plugging these numbers into the quadratic formula produces solutions, as follows:

$$x = \{4 \pm [(-4)^2 - (4 \times -3 \times -2)]^{1/2}\}/(2 \times -3)$$
$$= 4 \pm (16 - 24)^{1/2}/(-6)$$
$$= 4 \pm (-8)^{1/2}/(-6)$$

We are confronted with the square root of -8 in the solution. What is this? It isn't a real number, but how can we ignore it? We'll look into this matter shortly. For now, let's work out a couple of problems with quadratics involving only the real numbers.

THE PARABOLA

When we substitute y for 0 in the standard form of a quadratic equation, and then graph the resulting relation with x on the horizontal axis and y on the vertical axis, we get a curve called a *parabola*.

Suppose we want to plot a graph of the following equation:

$$ax^2 + bx + c = y$$

First, we determine the coordinates of the following point (x_0, y_0):

$$x_0 = -b/(2a)$$
$$y_0 = c - b^2/(4a)$$

This point represents the *base point* of the parabola; that is, the point at which the curvature is sharpest. Once this point is known, we must find four more points by "plugging in" values of x somewhat greater than and less than x_0, and then determining the corresponding y-values. These x-values, call them x_{-2}, x_{-1}, x_1, and x_2, should be equally spaced on either side of x_0, such that:

$$x_{-2} < x_{-1} < x_0 < x_1 < x_2$$
$$x_{-1} - x_{-2} = x_0 - x_{-1} = x_1 - x_0 = x_2 - x_1$$

This gives us five points that lie along the parabola, and that are symmetrical relative to the axis of the curve. The graph can then be inferred if the points are wisely chosen. Some trial and error might be required. If $a > 0$, the parabola opens upward. If $a < 0$, the parabola opens downward.

Plotting a parabola

Consider the following formula:

$$y = x^2 + 2x + 1$$

Using the above formula to calculate the base point, we get this:

$$x_0 = -2/2 = -1$$
$$y_0 = 1 - 4/4 = 1 - 1 = 0$$

Therefore $(x_0, y_0) = (-1, 0)$

We plot this point first, as shown in Fig. 6-1. Next, we plot points spaced at 1-unit horizontal-axis intervals on either side of x_0, as follows:

$$x_{-2} = x_0 - 2 = -3$$
$$y_{-2} = (-3)^2 + 2 \times (-3) + 1 = 9 - 6 + 1 = 4$$

Therefore $(x_{-2}, y_{-2}) = (-3, 4)$

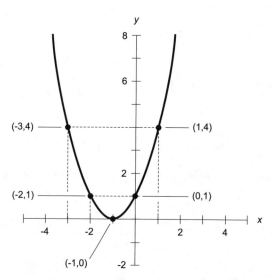

Fig. 6-1.   Graph of the quadratic equation $y = x^2 + 2x + 1$.

$$x_{-1} = x_0 - 1 = -2$$

$$y_{-1} = (-2)^2 + 2 \times (-2) + 1 = 4 - 4 + 1 = 1$$

Therefore $(x_{-1}, y_{-1}) = (-2, 1)$

$$x_1 = x_0 + 1 = 0$$

$$y_1 = 0^2 + 2 \times 0 + 1 = 0 + 0 + 1 = 1$$

Therefore $(x_1, y_1) = (0, 1)$

$$x_2 = x_0 + 2 = 1$$

$$y_2 = 1^2 + 2 \times 1 + 1 = 1 + 2 + 1 = 4$$

Therefore $(x_2, y_2) = (1, 4)$

From these five points, the parabola can be inferred. It is shown as a solid, heavy curve in Fig. 6-1.

Plotting another parabola

Let's try another example, this time with a parabola that opens downward. Consider the following formula:

$$y = -2x^2 + 4x - 5$$

The base point is:

$$x_0 = -4/-4 = 1$$

$$y_0 = -5 - 16/(-8) = -5 + 2 = -3$$

Therefore $(x_0, y_0) = (1, -3)$

This point is plotted first, as shown in Fig. 6-2. Next, we plot the following points:

$$x_{-2} = x_0 - 2 = -1$$

$$y_{-2} = -2 \times (-1)^2 + 4 \times (-1) - 5 = -2 - 4 - 5 = -11$$

Therefore $(x_{-2}, y_{-2}) = (-1, -11)$

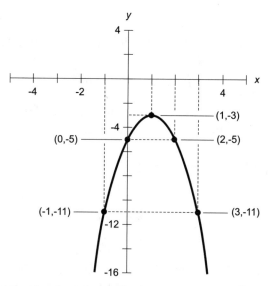

Fig. 6-2. Graph of the quadratic equation $y = -2x^2 + 4x - 5$.

$$x_{-1} = x_0 - 1 = 0$$

$$y_{-1} = -2 \times 0^2 + 4 \times 0 - 5 = -5$$

Therefore $(x_{-1}, y_{-1}) = (0, -5)$

$$x_1 = x_0 + 1 = 2$$

$$y_1 = -2 \times 2^2 + 4 \times 2 - 5 = -8 + 8 - 5 = -5$$

Therefore $(x_1, y_1) = (2, -5)$

$$x_2 = x_0 + 2 = 3$$

$$y_2 = -2 \times 3^2 + 4 \times 3 - 5 = -18 + 12 - 5 = -11$$

Therefore $(x_2, y_2) = (3, -11)$

From these five points, the parabola is inferred. It is the solid, heavy curve in Fig. 6-2.

PROBLEM 6-3

Find the real-number solution(s) to the following equation by factoring:

$$x^2 + 14x + 49 = 0$$

SOLUTION 6-3

Assuming this equation can be put into factored form, let's suppose the equation looks like this:

$$(x + t)(x + u) = 0$$

In this case, the solutions are $x = -t$ and $x = -u$. The trick is finding t and u such that we get the original equation when the factors are multiplied together. The above generalized equation looks like this when multiplied out:

$$(x + t)(x + u) = 0$$
$$x^2 + ux + tx + tu = 0$$
$$x^2 + (u + t)x + tu = 0$$

Are there any numbers t and u whose sum equals 14 and their product equals 49? It should not take you long to see that the answer is yes; this works out if $t = 7$ and $u = 7$. If we substitute these values into the generalized, factored quadratic and then multiply it out, we get the original equation that we want to solve:

$$(x + 7)(x + 7) = 0$$
$$x^2 + 7x + 7x + 49 = 0$$
$$x^2 + 14x + 49 = 0$$

Therefore, there is a single real-number solution to this equation: $x = -7$.

PROBLEM 6-4

Use the quadratic formula to solve the equation from the previous problem.

SOLUTION 6-4

Here is the equation in standard form again:

$$x^2 + 14x + 49 = 0$$

The coefficients from the general form, and which are defined in the quadratic formula above, are as follows:

$$a = 1$$
$$b = 14$$
$$c = 49$$

Here is what happens when we plug these coefficients into the quadratic formula:

$$x = \{-14 \pm [(14)^2 - (4 \times 1 \times 49)]^{1/2}\}/(2 \times 1)$$
$$= [-14 \pm (196 - 196)^{1/2}]/2$$
$$= (-14 \pm 0^{1/2})/2$$
$$= -14/2$$
$$= -7$$

Beyond Reality

Mathematicians symbolize the positive square root of -1, called the *unit imaginary number*, by using the lowercase italic letter i. Scientists and engineers symbolize it using the letter j, and henceforth, we will too.

IMAGINARY NUMBERS

Any imaginary number can be obtained by multiplying j by some real number q. The real number q is customarily written after j if q is positive or zero. If q happens to be a negative real number, then the absolute value of q is written after $-j$. Examples of imaginary numbers are $j3$, $-j5$, $j2.787$, and $-j\pi$.

The set of imaginary numbers can be depicted along a number line, just as can the real numbers. The so-called *imaginary number line* is usually shown standing on end, that is, vertically (Fig. 6-3). In a sense, the real-number line and the imaginary-number line are fraternal twins. As is the case with human twins, these two number lines, although they share similarities, are independent. The sets of imaginary and real numbers have one value, 0, in common. Thus:

$$j0 = 0$$

COMPLEX NUMBERS

A *complex number* consists of the sum of some real number and some imaginary number. The general form for a complex number k is:

$$k = p + jq$$

where p and q are real numbers.

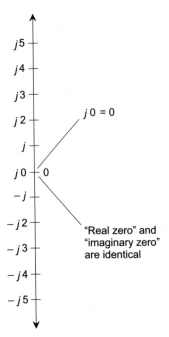

Fig. 6-3. The imaginary number line is just like the real number line, except that it is "stood on end" and all the quantities are multiplied by j.

Mathematicians, scientists, and engineers denote the set of complex numbers by placing the real-number and imaginary-number lines at right angles to each other, intersecting at the points on both lines corresponding to 0. The result is a rectangular coordinate plane (Fig. 6-4). Every point on this plane corresponds to a unique complex number, and every complex number corresponds to a unique point on the plane.

Now that you know a little about complex numbers, you might want to examine the solution to the following equation again:

$$-3x^2 - 4x - 2 = 0$$

Recall that the solution, derived using the quadratic formula, contains the quantity $(-8)^{1/2}$. An engineer or physicist would write this as $j8^{1/2}$, so the solution to the quadratic is:

$$x = (4 \pm j8^{1/2})/(-6)$$

This can be simplified to the standard form of a complex number, and then reduced to the lowest fractional form. Step-by-step, the simplification process

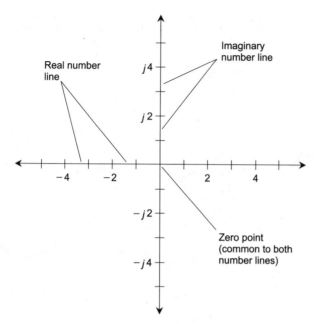

Fig. 6-4.   The complex number plane portrays real numbers on the horizontal axis and
imaginary numbers on the vertical axis.

goes like this:

$$x = (4 \pm j8^{1/2})/(-6)$$

$$= 4/6 \pm j[8^{1/2}/(-6)]$$

$$= 2/3 \pm j[2 \times 2^{1/2}/(-6)]$$

$$= 2/3 \pm j[2/(-6) \times 2^{1/2}]$$

$$= 2/3 \pm j(-1/3 \times 2^{1/2})$$

$$= 2/3 \pm [-j(1/3 \times 2^{1/2})]$$

$$= 2/3 \pm j(1/3 \times 2^{1/2})$$

This might not look any "simpler" at first glance, but it's good practice to
state complex numbers in standard form, and reduced to lowest fractions.
The last step, in which the minus sign disappears, is justified because adding
a negative is the same thing as subtracting a positive, and subtracting a
negative is the same thing as adding a positive.

The two complex solutions to the equation can be stated separately this way:

$$x = 2/3 + j(1/3 \times 2^{1/2})$$

or

$$x = 2/3 - j(1/3 \times 2^{1/2})$$

PROBLEM 6-5
Solve the following equation using the quadratic formula:

$$x^2 + 9 = 0$$

SOLUTION 6-5
In standard form showing all three coefficients a, b, and c, the equation looks like this:

$$1x^2 + 0x + 9 = 0$$

Thus the coefficients are:

$$a = 1$$
$$b = 0$$
$$c = 9$$

Plugging these numbers into the quadratic formula yields:

$$x = \{-0 \pm [0^2 - (4 \times 1 \times 9)]^{1/2}\}/(2 \times 1)$$
$$= \pm(-36)^{1/2}/2$$
$$= \pm j6/2$$
$$= \pm j3$$

PROBLEM 6-6
Write out the quadratic equation from the preceding problem in factored form.

SOLUTION 6-6
This would be tricky if we didn't already know the solutions. But we do, so it's easy:

$$(x + j3)(x - j3) = 0$$

You can verify that this works by "plugging in" the solutions derived from the quadratic formula in the previous problem.

One-Variable, Higher-Order Equations

As the exponents in single-variable equations get bigger, finding the solutions becomes a more challenging business. In the olden days, a lot of insight, guesswork, and tedium was involved in solving such equations. Today, scientists have the help of computers, and when problems are encountered containing equations with variables raised to large powers, they just let a computer take over. The material here is presented so you won't be taken aback if you ever come across *one-variable, higher-order equations*.

THE CUBIC

A cubic equation, also called a *one-variable, third-order equation* or a *third-order equation in one variable*, can be written in the following standard form:

$$ax^3 + bx^2 + cx + d = 0$$

where a, b, c, and d are constants, x is the variable, and $a \neq 0$.

If you're lucky, you'll be able to reduce a cubic equation to factored form to find real-number solutions $-r$, $-s$, and $-t$:

$$(x + r)(x + s)(x + t) = 0$$

Don't count on being able to factor a cubic equation. Sometimes it's easy, but most of the time it is nigh impossible. There is a *cubic formula* that can be used in a manner similar to the way in which the quadratic formula is used for quadratic equations, but it's complicated, and is beyond the scope of this discussion.

Plotting cubics

When we substitute y for 0 in the standard form of a cubic equation and then graph the resulting relation with x on the horizontal axis and y on the vertical axis, a curve with a characteristic shape results.

Figure 6-5 is a graph of the simplest possible cubic equation:

$$x^3 = y$$

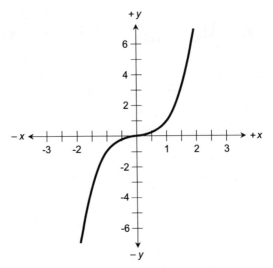

Fig. 6-5.   Graph of the cubic equation $x^3 = y$.

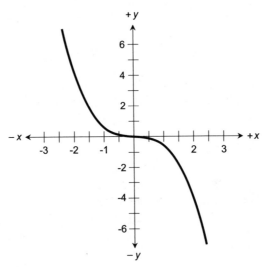

Fig. 6-6.   Graph of the cubic equation $(-1/2)x^3 = y$.

The domain and range of this function both encompass the entire set of real numbers. This makes the cubic curve different from the parabola, whose range is always limited to only a portion of the set of real numbers. Figure 6-6 is a graph of another simple cubic:

$$(-1/2)x^3 = y$$

In this case, the domain and range also span all the real numbers.

The *inflection point*, or the place where the curve goes from concave downward to concave upward, is at the origin $(0,0)$ in both Fig. 6-5 and Fig. 6-6. This is because the coefficients b, c, and d are all equal to 0. When these coefficients are nonzero, the inflection point is not necessarily at the origin.

THE QUARTIC

A *quartic equation*, also called a *one-variable, fourth-order equation* or a *fourth-order equation in one variable*, can be written in the following standard form:

$$ax^4 + bx^3 + cx^2 + dx + e = 0$$

where a, b, c, d, and e are constants, x is the variable, and $a \neq 0$.

Once in a while you will be able to reduce a quartic equation to factored form to find real-number solutions $-r$, $-s$, $-t$, and $-u$:

$$(x + r)(x + s)(x + t)(x + u) = 0$$

As is the case with the cubic, you will be lucky if you can factor a quartic equation into this form and thus find four real-number solutions with ease.

Plotting quartics

When we substitute y for 0 in the standard form of a quartic equation and then graph the resulting relation with x on the horizontal axis and y on the vertical axis, a curve with a "parabola-like" shape is the result. But this curve is not a true parabola. It is distorted – flattened down in a sense – as can be seen when it is compared with a true parabola. The bends in the curve are sharper than those in the parabola, and the curve "takes off" more steeply beyond the bends.

Figure 6-7 is a graph of the simplest possible quartic equation:

$$x^4 = y$$

The domain encompasses the whole set of real numbers, but the range of this function spans only the non-negative real numbers. Figure 6-8 is a graph of another simple quartic:

$$(-1/2)x^4 = y$$

In this case, the domain spans all the real numbers, but the range spans only the non-positive real numbers.

You are welcome to "play around" with various quartic equations to see what their graphs look like. Their shapes, positions, and orientations can vary depending on the values of the coefficients b, c, d, and e.

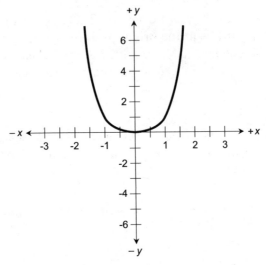

Fig. 6-7. Graph of the quartic equation $x^4 = y$.

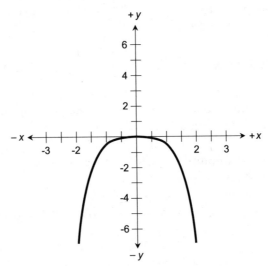

Fig. 6-8. Graph of the quartic equation $(-1/2)x^4 = y$.

THE QUINTIC

A *quintic equation*, also called a *one-variable, fifth-order equation* or a *fifth-order equation in one variable*, can be written in the following standard form:

$$ax^5 + bx^4 + cx^3 + dx^2 + ex + f = 0$$

where a, b, c, d, e, and f are constants, x is the variable, and $a \neq 0$.

There is a remote possibility that, if you come across a quintic, you'll be able to reduce it to factored form to find real-number solutions $-r$, $-s$, $-t$, $-u$, and $-v$:

$$(x + r)(x + s)(x + t)(x + u)(x + v) = 0$$

As is the case with the cubic and the quartic, you will be lucky if you can factor a quintic equation into this form.

Plotting quintics

When we substitute y for 0 in the standard form of a quintic equation and then graph the resulting relation with x on the horizontal axis and y on the vertical axis, the resulting curve looks something like that of the graph of the cubic equation. The difference is that the bends are sharper, and the graphs run off more steeply toward "positive infinity" and "negative infinity" beyond the bends.

Figure 6-9 is a graph of the simplest possible quintic equation:

$$x^5 = y$$

The domain and range of this function both span the set of reals. Figure 6-10 is a graph of another simple quintic:

$$(-1/2)x^5 = y$$

In this case, the domain and range also span all the real numbers.

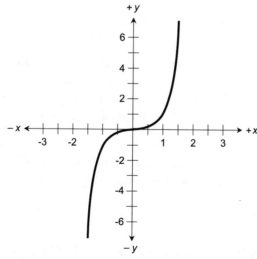

Fig. 6-9. Graph of the quintic equation $x^5 = y$.

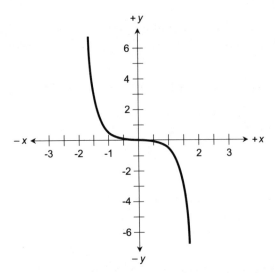

Fig. 6-10.   Graph of the quintic equation $(-1/2)x^5 = y$.

The inflection point for the graph of the quintic is at the origin in both Fig. 6-9 and Fig. 6-10. As in the case of equations of lower order that are shown in this chapter, the reason is that the coefficients other than a are all equal to 0.

THE nth-ORDER EQUATION

A *one-variable, nth-order equation* can be written in the following standard form:

$$a_1x^n + a_2x^{n-1} + a_3x^{n-2} + \cdots + a_{n-2}x^2 + a_{n-1}x + a_n = 0$$

where $a_1, a_2, \ldots, a_n$ are constants, x is the variable, and $a_1 \neq 0$. We won't even think about trying to factor an equation like this in general, although specific cases might lend themselves to factorization. Solving nth-order equations, where $n > 5$, practically demands the use of a computer.

PROBLEM 6-7
What happens to the general shape of the graph of the following equation if n is a positive even integer and n becomes larger without limit?

$$x^n = y$$

SOLUTION 6-7
If $x = 0$, then $y = 0$. If $x = -1$ or $x = 1$, then $y = 1$. These two facts are true no matter what the value of n, as long as it is even. As n increases, the curve

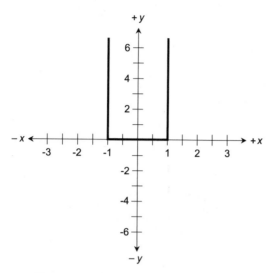

Fig. 6-11. Illustration for Problem 6-7.

tends more and more toward 0 from the positive side when $-1 < x < 0$ and when $0 < x < 1$. Also, the curve rises ever-more-steeply toward "positive infinity" $(+\infty)$ when $x < -1$ or $x > 1$. Figure 6-11 is an approximate drawing of the graph of the equation $x^{998} = y$. The bends are not perfectly squared-off, but they're pretty close. The curve "takes off" in almost, but not perfectly, vertical directions.

PROBLEM 6-8
What happens to the general shape of the graph of the following equation if n is a positive odd integer and n becomes larger without limit?

$$x^n = y$$

SOLUTION 6-8
If $x = 0$, then $y = 0$. If $x = -1$, then $y = -1$. If $x = 1$, then $y = 1$. These two facts are true regardless of the value of n, as long as it is odd. As n increases, the curve tends more and more toward 0 from the negative side when $-1 < x < 0$, and more and more toward 0 from the positive side when $0 < x < 1$. The curve descends ever-more-steeply toward "negative infinity" $(-\infty)$ when $x < -1$, and rises ever-more-steeply toward "positive infinity" $(+\infty)$ when $x > 1$. Figure 6-12 is an approximate drawing of the graph of the equation $x^{999} = y$. As with the curve shown in Fig. 6-11, the "take-off" slopes are almost, but not quite, vertical.

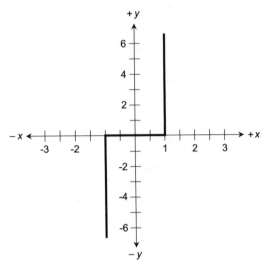

Fig. 6-12.   Illustration for Problem 6-8.

Quiz

Refer to the text in this chapter if necessary. A good score is eight correct. Answers are in the back of the book.

1. A point on the graph of a function where the curve goes from concave upward to concave downward is called
    (a) an inflection point
    (b) a focal point
    (c) a base point
    (d) a slope point

2. The general form of a quadratic equation is:

    $$ax^2 + bx + c = 0$$

    Suppose a is a negative real number, b and c are real numbers, and $(b^2 - 4ac) = 0$. Then which of the following statements is true?
    (a) The equation has a real-number solution.
    (b) The equation has two different real-number solutions.
    (c) The equation has no real-number solutions.
    (d) The equation has no solutions at all.

3. The equation $x^5 = 3x^2 + 2$ is an example of
    (a) a second-order equation

    (b) a third-order equation
    (c) a fourth-order equation
    (d) a fifth-order equation

4.   Which of the following equations has a graph that is not a parabola?
    (a) $x^2 = y$
    (b) $-x^2 + 2x = y$
    (c) $5x^2 - 2x + 4 = y$
    (d) $4x + 5 = y$

5.   In standard form, the factored equation $(x - 10)(x + 10) = 0$ is:
    (a) $x^2 + 10x + 100 = 0$
    (b) $x^2 - 10x - 100 = 0$
    (c) $x^2 + 100 = 0$
    (d) $x^2 - 100 = 0$

6.   What is, or are, the real-number solutions of the equation $(x - 10)$ $(x + 10) = 0$?
    (a) -10
    (b) 10
    (c) -10 and 10
    (d) There are none.

7.   Which of the following numbers, when multiplied by itself, is equal to -100?
    (a) -10
    (b) 10
    (c) $j10$
    (d) $10 + j10$

8.   The domain of the function $2x^3 - x^2 + 4x + 2 = y$ consists of
    (a) the set of real numbers greater than 2
    (b) the set of real numbers smaller than 2
    (c) the set of real numbers greater than or equal to 0
    (d) the entire set of real numbers

9.   Consider the following equation:

$$(x - 6)(x + j10)(x - 14)(x + j2) = 0$$

Which of the following numbers is not a solution to this equation?
    (a) 6
    (b) 14
    (c) $j2$
    (d) $-j10$

10. The solution(s) to the equation $x^2 - 2 = 0$ is or are
    (a) imaginary
    (b) real, but irrational
    (c) real and rational
    (d) nonexistent

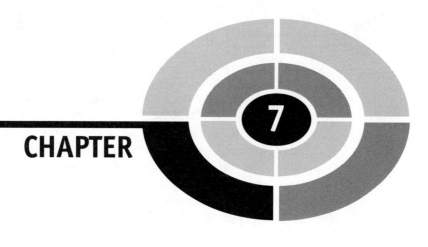

CHAPTER

A Statistics Sampler

If you want to understand anything about a scientific discipline, you must know the terminology. Statistics is no exception. This chapter defines some of the most common terms used in statistics.

Experiments and Variables

Statistics is the analysis of information. In particular, statistics is concerned with *data*: information expressed as measurable or observable quantities. Statistical data is usually obtained by looking at the real world or universe, although it can also be generated by computers in "artificial worlds."

EXPERIMENT

In statistics, an *experiment* is an act of collecting data with the intent of learning or discovering something. For example, we might conduct an experiment to determine the most popular channels for frequency-modulation (FM) radio broadcast stations whose transmitters are located in American towns having less than 5000 people. Or we might conduct an experiment to determine the lowest barometric pressures inside the eyes of all the Atlantic hurricanes that take place during the next 10 years.

Experiments often, but not always, require specialized instruments to measure quantities. If we conduct an experiment to figure out the average test scores of high-school seniors in Wyoming who take a certain standardized test at the end of this school year, the only things we need are the time, energy, and willingness to collect the data. But a measurement of the minimum pressure inside the eye of a hurricane requires sophisticated hardware in addition to time, energy, and courage.

VARIABLE (IN GENERAL)

In mathematics, a *variable*, also called an *unknown*, is a quantity whose value is not necessarily specified, but that can be determined according to certain rules. Mathematical variables are expressed using italicized letters of the alphabet, usually in lowercase. For example, in the expression $x + y + z = 5$, the letters x, y, and z are variables that represent numbers.

In statistics, variables are similar to those in mathematics. But there are some subtle distinctions. Perhaps most important is this: In statistics, a variable is always associated with one or more experiments.

DISCRETE VARIABLE

In statistics, a *discrete variable* is a variable that can attain only specific values. The number of possible values is countable. Discrete variables are like the channels of a television set or digital broadcast receiver. It's easy to express the value of a discrete variable, because it can be assumed exact.

When a disc jockey says "This is radio 97.1," it means that the assigned channel center is at a frequency of 97.1 megahertz, where a megahertz (MHz) represents a million cycles per second. The assigned channels in the FM broadcast band are separated by an *increment* (minimum difference) of 0.2 MHz. The next lower channel from 97.1 MHz is at 96.9 MHz, and the next higher one is at 97.3 MHz. There is no "in between." No two channels can be closer together than 0.2 MHz in the set of assigned standard FM broadcast channels in the United States. The lowest channel is at 88.1 MHz and the highest is at 107.9 MHz (Fig. 7-1).

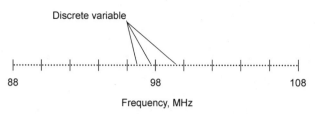

Fig. 7-1.   The individual channels in the FM broadcast band constitute values of a
discrete variable.

Other examples of discrete variables are:

- The number of people voting for each of the various candidates in a political election.
- The scores of students on a standardized test (expressed as a percentage of correct answers).
- The number of car drivers caught speeding every day in a certain town.
- The earned-run averages of pitchers in a baseball league (in runs per 9 innings or 27 outs).

All these quantities can be expressed as exact values.

CONTINUOUS VARIABLE

A *continuous variable* can attain infinitely many values over a certain span or range. Instead of existing as specific values in which there is an increment between any two, a continuous variable can change value to an arbitrarily tiny extent.

Continuous variables are something like the set of radio frequencies to which an analog FM broadcast receiver can be tuned. The radio frequency is adjustable continuously, say from 88 MHz to 108 MHz for an FM headset receiver with analog tuning (Fig. 7-2). If you move the tuning dial a little, you can make the received radio frequency change by something less than 0.2 MHz, the separation between adjacent assigned transmitter channels. There is no limit to how small the increment can get. If you have a light enough touch, you can adjust the received radio frequency by 0.02 MHz, or 0.002 MHz, or even 0.000002 MHz.

Other examples of continuous variables are:

- Temperature in degrees Celsius.
- Barometric pressure in millibars.
- Brightness of a light source in candela.
- Intensity of the sound from a loudspeaker in decibels with respect to the threshold of hearing.

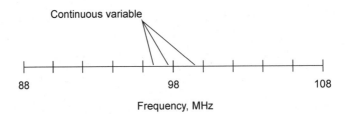

Fig. 7-2. The frequency to which an analog FM broadcast receiver can be set is an example of a continuous variable.

Such quantities can never be determined exactly. There is always some instrument or observation error, even if that error is so small that it does not have a practical effect on the outcome of an experiment.

Populations and Samples

In statistics, the term *population* refers to a particular set of items, objects, phenomena, or people being analyzed. These items, also called *elements*, can be actual subjects such as people or animals, but they can also be numbers or definable quantities expressed in physical units.

Consistent with the above definitions of variables, some examples of populations are as follows:

- Assigned radio frequencies (in megahertz) of all FM broadcast transmitters in the United States.
- Temperature readings (in degrees Celsius) at hourly intervals last Wednesday at various locations around the city of New York.
- Minimum barometric-pressure levels (in millibars) at the centers of all the hurricanes in recorded history.
- Brightness levels (in candela) of all the light bulbs in offices in Minneapolis.
- Sound-intensity levels (in decibels relative to the threshold of hearing) of all the electric vacuum cleaners in the world.

SAMPLE, EVENT, AND CENSUS

A *sample* of a population is a subset of that population. It can be a set consisting of only one value, reading, or measurement singled out from a population, or it can be a subset that is identified according to certain characteristics. The physical unit (if any) that defines a sample is always the same as the physical unit that defines the main, or parent, population. A single element of a sample is called an *event*.

Consistent with the above definitions of variables, some samples are:

- Assigned radio frequencies of FM broadcast stations whose transmitters are located in the state of Ohio.
- Temperature readings at 1:00 P.M. local time last Wednesday at various locations around the city of New York.
- Minimum barometric-pressure levels at the centers of Atlantic hurricanes during the decade 1991–2000.
- Brightness levels of halogen bulbs in offices in Minneapolis.
- Sound-intensity levels of the electric vacuum cleaners used in all the households in Rochester, Minnesota.

When a sample consists of the whole population, it is called a *census*. When a sample consists of a subset of a population whose elements are chosen at random, it is called a *random sample*.

RANDOM VARIABLE

A *random variable* is a discrete or continuous variable whose value cannot be predicted in any given instance. Such a variable is usually defined within a certain range of values, such as 1 through 6 in the case of a thrown die, or from 88 MHz to 108 MHz in the case of an FM broadcast channel.

It is often possible to say, in a given scenario, that some values of a random variable are more likely to turn up than others. In the case of a thrown die, assuming the die is not "weighted," all of the values 1 through 6 are equally likely to turn up. When considering the FM broadcast channels of public radio stations, it is tempting to suppose (but this would have to be confirmed by observation) that transmissions are made more often at the lower radio-frequency range than at the higher range. Perhaps you have noticed that there is a greater concentration of public radio stations in the 4-MHz-wide sample from 88 MHz to 92 MHz than in, say, the equally wide sample from 100 MHz to 104 MHz.

In order for a variable to be random, the only requirement is that it be impossible to predict its value in any single instance. If you contemplate throwing a die one time, you can't predict how it will turn up. If you contemplate throwing a dart one time at a map of the United States while wearing a blindfold, you have no way of knowing, in advance, the lowest radio frequency of all the FM broadcast stations in the town nearest the point where the dart will hit.

FREQUENCY

The *frequency* of a particular outcome (result) of an event is the number of times that outcome occurs within a specific sample of a population. Don't confuse this with radio broadcast or computer processor frequencies! In statistics, the term "frequency" means "often-ness." There are two species of statistical frequency: *absolute frequency* and *relative frequency*.

Suppose you toss a die 6000 times. If the die is not "weighted," you should expect that the die will turn up showing one dot approximately 1000 times, two dots approximately 1000 times, and so on, up to six dots approximately 1000 times. The absolute frequency in such an experiment is therefore approximately 1000 for each face of the die. The relative frequency for each of the six faces is approximately 1 in 6, which is equivalent to about 16.67%.

PARAMETER

A specific, well-defined characteristic of a population is known as a *parameter* of that population. We might want to know such parameters as the following, concerning the populations mentioned above:

- The most popular assigned FM broadcast frequency in the United States.
- The highest temperature reading in the city of New York as determined at hourly intervals last Wednesday.
- The average minimum barometric-pressure level or measurement at the centers of all the hurricanes in recorded history.
- The lowest brightness level found in all the light bulbs in offices in Minneapolis.
- The highest sound-intensity level found in all the electric vacuum cleaners used in the world.

STATISTIC

A specific characteristic of a sample is called a *statistic* of that sample. We might want to know such statistics as these, concerning the samples mentioned above:

- The most popular assigned frequency for FM broadcast stations in Ohio.
- The highest temperature reading at 1:00 P.M. local time last Wednesday in New York.

- The average minimum barometric-pressure level or measurement at the centers of Atlantic hurricanes during the decade 1991–2000.
- The lowest brightness level found in all the halogen bulbs in offices in Minneapolis.
- The highest sound-intensity level found in electric vacuum cleaners used in households in Rochester, Minnesota.

Distributions

A *distribution* is a description of the set of possible values that a random variable can take. This can be done by noting the absolute or relative frequency. A distribution can be illustrated in terms of a table, or in terms of a graph.

DISCRETE VERSUS CONTINUOUS

Table 7-1 shows the results of a single, hypothetical experiment in which a die is tossed 6000 times. Figure 7-3 is a vertical bar graph showing the same data as Table 7-1. Both the table and the graph are distributions that describe the behavior of the die. If the experiment is repeated, the results will differ. If a huge number of experiments is carried out, assuming the die is not "weighted," the relative frequency of each face (number) turning up will approach 1 in 6, or approximately 16.67%.

Table 7-1   Results of a single, hypothetical experiment in which an "unweighted" die is tossed 6000 times.

Face of die	Number of times face turns up
1	968
2	1027
3	1018
4	996
5	1007
6	984

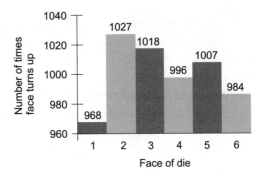

Fig. 7-3. Results of a single, hypothetical experiment in which an "unweighted" die is tossed 6000 times.

Table 7-2   Number of days on which measurable rain occurs in a specific year, in five hypothetical towns.

Town name	Number of days in year with measurable precipitation
Happyville	108
Joytown	86
Wonderdale	198
Sunnywater	259
Rainy Glen	18

Table 7-2 shows the number of days during the course of a 365-day year in which measurable precipitation occurs within the city limits of five different hypothetical towns. Figure 7-4 is a horizontal bar graph showing the same data as Table 7-2. Again, both the table and the graph are distributions. If the same experiment were carried out for several years in a row, the results would differ from year to year. Over a period of many years, the relative frequencies would converge towards certain values, although long-term climate change might have effects not predictable or knowable in our lifetimes.

Both of the preceding examples involve discrete variables. When a distribution is shown for a continuous variable, a graph must be used. Figure 7-5 is a distribution that denotes the relative amount of energy available from sunlight, per day during the course of a calendar year, at a hypothetical city in the northern hemisphere.

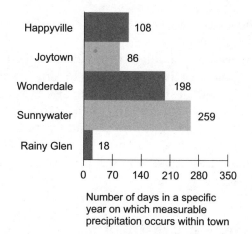

Fig. 7-4. Measurable precipitation during a hypothetical year, in five different imaginary towns.

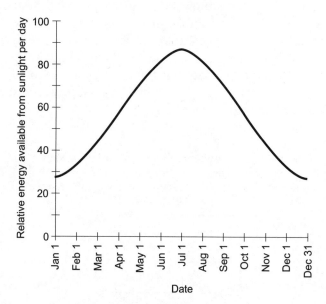

Fig. 7-5. Relative energy available from sunlight, per day, during the course of a calendar year at a hypothetical location.

FREQUENCY DISTRIBUTION

In both of the above examples (the first showing the results of 6000 die tosses and the second showing the days with precipitation in five hypothetical towns), the scenarios are portrayed with frequency as the dependent variable. This is true of the tables as well as the graphs. Whenever frequency is portrayed as the dependent variable in a distribution, that distribution is called a *frequency distribution*.

Suppose we complicate the situation involving dice. Instead of one person tossing one die 6000 times, we have five people tossing five different dice, and each person tosses the same die 6000 times. The dice are colored red, orange, yellow, green, and blue, and are manufactured by five different companies, called Corp. A, Corp. B, Corp. C, Corp. D, and Corp. E, respectively. Four of the dice are "weighted" and one is not. There are thus 30,000 die tosses to tabulate or graph in total. When we conduct this experiment, we can tabulate the data in at least two ways.

Ungrouped frequency distribution

The simplest way to tabulate the die toss results as a frequency distribution is to combine all the tosses and show the total frequency for each die face 1 through 6. A hypothetical example of this result, called an *ungrouped frequency distribution*, is shown in Table 7-3. We don't care about the weighting characteristics of each individual die, but only about potential biasing of the entire set. It appears that, for this particular set of dice, there is some bias in favor of faces 4 and 6, some bias against faces 1 and 3, and little or no bias either for or against faces 2 and 5.

Table 7-3 An ungrouped frequency distribution showing the results of a single, hypothetical experiment in which five different dice, some "weighted" and some not, are each tossed 6000 times.

Face of die	Toss results for all die
1	4857
2	4999
3	4626
4	5362
5	4947
6	5209

Grouped frequency distribution

If we want to be more detailed, we can tabulate the frequency for each die face 1 through 6 separately for each die. A hypothetical product of this effort, called a *grouped frequency distribution*, is shown in Table 7-4. The results are grouped according to manufacturer and die color. From this distribution, it is apparent that some of the dice are heavily "weighted." Only the green die, manufactured by Corp. D, seems to lack any bias. If you are astute, you will notice (or at least strongly suspect) that the green die here is the same die, with results gathered from the same experiment, as is portrayed in Table 7-1 and Fig. 7-3.

PROBLEM 7-1

Suppose you add up all the numbers in each column of Table 7-4. What should you expect, and why? What should you expect if the experiment is repeated many times?

SOLUTION 7-1

Each column should add up to 6000. This is the number of times each die (red, orange, yellow, green, or blue) is tossed in the experiment. If the sum of the numbers in any of the columns is not equal to 6000, then the

Table 7-4   A grouped frequency distribution showing the results of a single, hypothetical experiment in which five different dice, some "weighted" and some not, and manufactured by five different companies, are each tossed 6000 times.

Face of die	Toss results by manufacturer				
	Red Corp. A	Orange Corp. B	Yellow Corp. C	Green Corp. D	Blue Corp. E
1	625	1195	1689	968	380
2	903	1096	1705	1027	268
3	1300	890	1010	1018	408
4	1752	787	540	996	1287
5	577	1076	688	1007	1599
6	843	956	368	984	2058

experiment was done in a faulty way, or else there is an error in the compilation of the table. If the experiment is repeated many times, the sums of the numbers in each column should always be 6000.

PROBLEM 7-2
Suppose you add up all the numbers in each row of Table 7-4. What should you expect, and why? What should you expect if the experiment is repeated many times?

SOLUTION 7-2
The sums of the numbers in the rows will vary, depending on the bias of the set of dice considered as a whole. If, taken all together, the dice show any bias, and if the experiment is repeated many times, the sums of the numbers should be consistently lower for some rows than for other rows.

PROBLEM 7-3
Each small rectangle in a table, representing the intersection of one row with one column, is called a *cell* of the table. What do the individual numbers in the cells of Table 7-4 represent?

SOLUTION 7-3
The individual numbers are absolute frequencies. They represent the actual number of times a particular face of a particular die came up during the course of the experiment.

More Definitions

Here are some more definitions you should learn in order to get comfortable reading or talking about statistics.

CUMULATIVE ABSOLUTE FREQUENCY

When data are tabulated, the absolute frequencies are often shown in one or more columns. Look at Table 7-5, for example. This shows the results of the tosses of the blue die in the experiment we looked at a while ago. The first column shows the number on the die face. The second column shows the absolute frequency for each face, or the number of times each face turned up during the experiment. The third column shows the *cumulative absolute frequency*, which is the sum of all the absolute frequency values in table cells at or above the given position.

Table 7-5   Results of an experiment in which a "weighted" die is tossed 6000 times, showing absolute frequencies and cumulative absolute frequencies.

Face of die	Absolute frequency	Cumulative absolute frequency
1	380	380
2	268	648
3	408	1056
4	1287	2343
5	1599	3942
6	2058	6000

The cumulative absolute frequency numbers in a table always ascend (increase) as you go down the column. The highest cumulative absolute frequency value should be equal to the sum of all the individual absolute frequency numbers. In this instance, it is 6000, the number of times the blue die was tossed.

CUMULATIVE RELATIVE FREQUENCY

Relative frequency values can be added up down the columns of a table, in exactly the same way as the absolute frequency values are added up. When this is done, the resulting values, usually expressed as percentages, show the *cumulative relative frequency*.

Examine Table 7-6. This is a more detailed analysis of what happened with the blue die in the above-mentioned experiment. The first, second, and fourth columns in Table 7-6 are identical with the first, second, and third columns in Table 7-5. The third column in Table 7-6 shows the percentage represented by each absolute frequency number. These percentages are obtained by dividing the number in the second column by 6000, the total number of tosses. The fifth column shows the cumulative relative frequency, which is the sum of all the relative frequency values in table cells at or above the given position.

The cumulative relative frequency percentages in a table, like the cumulative absolute frequency numbers, always ascend as you go down

Table 7-6 Results of an experiment in which a "weighted" die is tossed 6000 times, showing absolute frequencies, relative frequencies, cumulative absolute frequencies, and cumulative relative frequencies.

Face of die	Absolute frequency	Relative frequency	Cumulative absolute frequency	Cumulative relative frequency
1	380	6.33%	380	6.33%
2	268	4.47%	648	10.80%
3	408	6.80%	1056	17.60%
4	1287	21.45%	2343	39.05%
5	1599	26.65%	3942	65.70%
6	2058	34.30%	6000	100.00%

the column. The total cumulative relative frequency should be equal to 100%. In this sense, the cumulative relative frequency column in a table can serve as a *checksum*, helping to ensure that the entries have been tabulated correctly.

MEAN

The *mean* for a discrete variable in a distribution is the mathematical average of all the values. If the variable is considered over the entire population, the average is called the *population mean*. If the variable is considered over a particular sample of a population, the average is called the *sample mean*. There can be only one population mean for a population, but there can be many different sample means. The mean is often denoted by the lowercase Greek letter mu, in italics (μ). Sometimes it is also denoted by an italicized lowercase English letter, usually x, with a bar (vinculum) over it.

Table 7-7 shows the results of a 10-question test, given to a class of 100 students. As you can see, every possible score is accounted for. There are some people who answered all 10 questions correctly; there are some who did not get a single answer right. In order to determine the mean score for the whole class on this test – that is, the population mean, called μ_p – we must add up the scores of each and every student, and then divide by 100.

Table 7-7   Scores on a 10-question test taken by 100 students.

Test score	Absolute frequency	Letter grade
10	5	A
9	6	A
8	19	B
7	17	B
6	18	C
5	11	C
4	6	D
3	4	D
2	4	F
1	7	F
0	3	F

First, let's sum the products of the numbers in the first and second columns. This will give us 100 times the population mean:

$$(10 \times 5) + (9 \times 6) + (8 \times 19) + (7 \times 17) + (6 \times 18) + (5 \times 11)$$
$$+ (4 \times 6) + (3 \times 4) + (2 \times 4) + (1 \times 7) + (0 \times 3)$$
$$= 50 + 54 + 152 + 119 + 108 + 55 + 24 + 12 + 8 + 7 + 0$$
$$= 589$$

Dividing this by 100, the total number of test scores (one for each student who turns in a paper), we obtain $\mu_p = 589/100 = 5.89$.

The teacher in this class has assigned letter grades to each score. Students who scored 9 or 10 correct received grades of A; students who got scores of 7 or 8 received grades of B; those who got scores of 5 or 6 got grades of C; those who got scores of 3 or 4 got grades of D; those who got less than 3 correct answers received grades of F. The assignment of grades, informally known as the "curve," is a matter of teacher temperament and doubtless would seem arbitrary to the students who took this test. (Some people might consider the "curve" in this case to be overly lenient, while a few might think it is too severe.)

PROBLEM 7-4

What are the sample means for each grade in the test whose results are tabulated in Table 7-7? Use rounding to determine the answers to two decimal places.

SOLUTION 7-4

Let's call the sample means μ_{sa} for the grade of A, μ_{sb} for the grade of B, and so on down to μ_{sf} for the grade of F.

To calculate μ_{sa}, note that 5 students received scores of 10, while 6 students got scores of 9, both scores good enough for an A. This is a total of $5 + 6$, or 11, students getting the grade of A. Therefore:

$$\mu_{sa} = [(5 \times 10) + (6 \times 9)]/11$$

$$= (50 + 54)/11$$

$$= 104/11$$

$$= 9.45$$

To find μ_{sb}, observe that 19 students scored 8, and 17 students scored 7. Thus, $19 + 17$, or 36, students received grades of B. Calculating:

$$\mu_{sb} = [(19 \times 8) + (17 \times 7)]/36$$

$$= (152 + 119)/36$$

$$= 271/36$$

$$= 7.53$$

To determine μ_{sc}, check the table to see that 18 students scored 6, while 11 students scored 5. Therefore, $18 + 11$, or 29, students did well enough for a C. Grinding out the numbers yields this result:

$$\mu_{sc} = [(18 \times 6) + (11 \times 5)]/29$$

$$= (108 + 55)/29$$

$$= 163/29$$

$$= 5.62$$

To calculate μ_{sd}, note that 6 students scored 4, while 4 students scored 3. This means that $6 + 4$, or 10, students got grades of D:

$$\mu_{sd} = [(6 \times 4) + (4 \times 3)]/10$$
$$= (24 + 12)/10$$
$$= 36/10$$
$$= 3.60$$

Finally, we determine μ_{sf}. Observe that 4 students got scores of 2, 7 students got scores of 1, and 3 students got scores of 0. Thus, $4 + 7 + 3$, or 14, students failed the test:

$$\mu_{sf} = [(4 \times 2) + (7 \times 1) + (3 \times 0)]/14$$
$$= (8 + 7 + 0)/14$$
$$= 15/14$$
$$= 1.07$$

MEDIAN

If the number of elements in a distribution is even, then the *median* is the value such that half the elements have values greater than or equal to it, and half the elements have values less than or equal to it. If the number of elements is odd, then the median is the value such that the number of elements having values greater than or equal to it is the same as the number of elements having values less than or equal to it. The word "median" is synonymous with "middle."

Table 7-8 shows the results of the 10-question test described above, but instead of showing letter grades in the third column, the cumulative absolute frequency is shown instead. The tally is begun with the top-scoring papers and proceeds in order downward. (It could just as well be done the other way, starting with the lowest-scoring papers and proceeding upward.) When the scores of all 100 individual papers are tallied this way, so they are in order, the scores of the 50th and 51st papers – the two in the middle – are found to be 6 correct. Thus, the median score is 6, because half the students scored 6 or above, and the other half scored 6 or below.

It's possible that in another group of 100 students taking this same test, the 50th paper would have a score of 6 while the 51st paper would have a score of 5. When two values "compete," the median is equal to their average. In this case it would be midway between 5 and 6, or 5.5.

Table 7-8 The median can be determined by tabulating the cumulative absolute frequencies.

Test score	Absolute frequency	Cumulative absolute frequency
10	5	5
9	6	11
8	19	30
7	17	47
6 (partial)	3	50
6 (partial)	15	65
5	11	76
4	6	82
3	4	86
2	4	90
1	7	97
0	3	100

MODE

The *mode* for a discrete variable is the value that occurs the most often. In the test whose results are shown in Table 7-7, the most "popular" or often-occurring score is 8 correct answers. There were 19 papers with this score. No other score had that many results. Therefore, the mode in this case is 8.

Suppose that another group of students took this test, and there were two scores that occurred equally often? For example, suppose 16 students got 8 answers right, and 16 students also got 6 answers right? In this case there are two modes: 6 and 8. This distribution would be called a *bimodal distribution*. It's even possible that three scores would be equally popular, resulting in a *trimodal distribution*.

Now imagine there are only 99 students in a class, and there are exactly 9 students who get each of the 11 possible scores (from 0 to 10 correct answers).

In this distribution, there is no mode. Or, we might say, the mode is not defined.

PROBLEM 7-5
Draw a vertical bar graph showing all the absolute-frequency data from Table 7-5, the results of a "weighted" die-tossing experiment. Portray each die face on the horizontal axis. Let light gray vertical bars show the absolute frequency numbers, and let dark gray vertical bars show the cumulative absolute frequency numbers.

SOLUTION 7-5
Figure 7-6 shows such a graph. The numerical data is not listed at the tops of the bars in order to avoid excessive clutter.

PROBLEM 7-6
Draw a horizontal bar graph showing all the relative-frequency data from Table 7-6, another portrayal of the results of a "weighted" die-tossing experiment. Show each die face on the vertical axis. Let light gray horizontal bars show the relative frequency percentages, and dark gray horizontal bars show the cumulative relative frequency percentages.

SOLUTION 7-6
Figure 7-7 is an example of such a graph. Again, the numerical data is not listed at the ends of the bars, in the interest of neatness.

PROBLEM 7-7
Draw a point-to-point graph showing the absolute frequencies of the 10-question test described by Table 7-7. Mark the population mean, the median, and the mode with distinctive vertical lines, and label them.

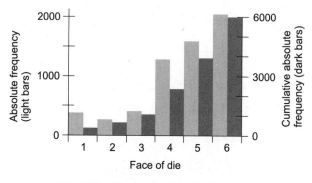

Fig. 7-6.   Illustration for Problem 7-5.

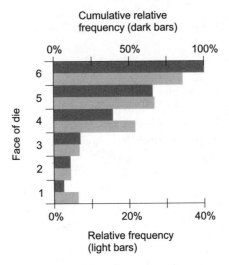

Fig. 7-7.   Illustration for Problem 7-6.

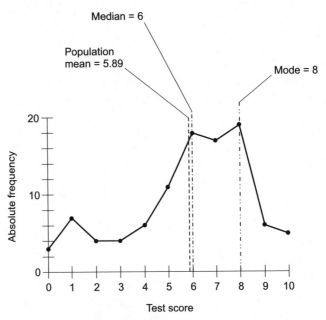

Fig. 7-8.   Illustration for Problem 7-7.

SOLUTION 7-7

Figure 7-8 is an example of such a graph. Numerical data is included for the population mean, median, and mode.

Quiz

Refer to the text in this chapter if necessary. A good score is eight correct. Answers are in the back of the book.

1. Suppose a large number of people take a 200-question test, and every single student gets exactly 100 answers right. In this case, the mode is
    (a) equal to 0
    (b) equal to 200
    (c) equal to the mean
    (d) undefined

2. In a frequency distribution:
    (a) the frequency is the dependent variable
    (b) the median is always equal to the mean
    (c) the mode represents the average value
    (d) the mean is always equal to the mode

3. A tabulation of cumulative absolute frequency values is useful in determining
    (a) the mode
    (b) the dependent variable
    (c) the median
    (d) the mean

4. A subset of a population is known as
    (a) a sample
    (b) a continuous variable
    (c) a random variable
    (d) a discrete variable

5. Imagine that 11 people take a 10-question test. Suppose one student gets 10 correct answers, one gets 9 correct, one gets 8 correct, and so on, all the way down to one student getting none correct. The mean, accurate to three decimal places, is
    (a) 4.545
    (b) 5.000
    (c) 5.500
    (d) undefined

6. Imagine that 11 people take a 10-question test. Suppose one student gets 10 correct answers, one gets 9 correct, one gets 8 correct, and

so on, all the way down to one student getting none correct. The median, accurate to three decimal places, is

(a) 4.545

(b) 5.000

(c) 5.500

(d) undefined

7. Imagine that 11 people take a 10-question test. Suppose one student gets 10 correct answers, one gets 9 correct, one gets 8 correct, and so on, all the way down to one student getting none correct. The mode, accurate to three decimal places, is

(a) 4.545

(b) 5.000

(c) 5.500

(d) undefined

8. Suppose a variable λ (lambda, pronounced "LAM-da," a lowercase Greek letter commonly used in physics and engineering to represent wavelength) can attain a value equal to any positive real number. In this instance, λ is an example of

(a) a continuous variable

(b) a discrete variable

(c) an absolute variable

(d) a relative variable

9. The largest cumulative absolute frequency in a set of numbers is equal to

(a) the sum of all the individual absolute frequency values

(b) twice the mean

(c) twice the median

(d) twice the mode

10. Which of the following is an expression of the most often-occurring value in a set of values?

(a) The average.

(b) The mode.

(c) The median.

(d) None of the above.

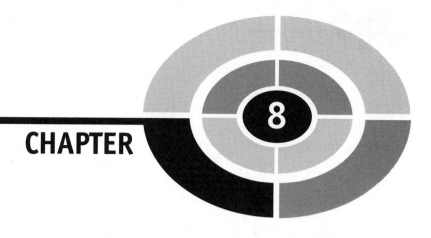

CHAPTER

8

Taking Chances

Probability is the proportion or percentage of the time that specified things happen. The term *probability* is also used in reference to the art and science of determining the proportion or percentage of the time that specified things happen.

The Probability Fallacy

We say something is true because we've seen or deduced it. If we believe something is true or has taken place but we aren't sure, it's tempting to say it is or was "likely." It's wise to resist this temptation.

BELIEF

When people formulate a theory, they often say that something "probably" happened in the distant past, or that something "might" exist somewhere, as-yet undiscovered, at this moment. Have you ever heard that there is a "good chance" that extraterrestrial life exists? Such a statement is meaningless. Either it exists, or it does not.

If you say "I believe the universe began with an explosion," you are stating the fact that you believe it, not the fact that it is true or that it is "probably" true. If you say "The universe began with an explosion!" your statement is logically sound, but it is a statement of a theory, not a proven fact. If you say "The universe probably started with an explosion," you are in effect suggesting that there were multiple pasts and the universe had an explosive origin in more than half of them. This is an instance of what can be called the *probability fallacy* (abbreviated PF), wherein probability is injected into a discussion inappropriately.

Whatever is, is. Whatever is not, is not. Whatever was, was. Whatever was not, was not. Either the universe started with an explosion, or it didn't. Either there is life on some other world, or there isn't.

PARALLEL WORLDS?

If we say that the "probability" of life existing elsewhere in the cosmos is 20%, we are in effect saying, "Out of n observed universes, where n is some large number, $0.2n$ universes have been found to have extraterrestrial life." That doesn't mean anything to those of us who have seen only one universe!

It is worthy of note that there are theories involving so-called *fuzzy truth*, in which some things "sort of happen." These theories involve degrees of truth that span a range over which probabilities can be assigned to occurrences in the past and present. An example of this is *quantum mechanics*, which is concerned with the behavior of subatomic particles. Quantum mechanics can get so bizarre that some scientists say, "If you claim to understand this stuff, then you are lying." We aren't going to deal with anything that esoteric.

WE MUST OBSERVE

Probability is usually defined according to the results of observations, although it is sometimes defined on the basis of theory alone. When the notion of probability is abused, seemingly sound reasoning can be employed to come to absurd conclusions. This sort of thing is done in industry every day, especially when the intent is to get somebody to do something that

will cause somebody else to make money. Keep your "probability fallacy radar" on when navigating through the real world.

If you come across an instance where an author (including me) says that something "probably happened," "is probably true," or "is likely to take place," think of it as another way of saying that the author believes or suspects that something happened, is true, or is expected to take place on the basis of experimentation or observation.

Definitions

Here are definitions of some common terms that will help us understand what we are talking about when we refer to probability.

EVENT VERSUS OUTCOME

The terms *event* and *outcome* are easily confused. An event is a single occurrence or trial in the course of an experiment. An outcome is the result of an event.

If you toss a coin 100 times, there are 100 separate events. Each event is a single toss of the coin. If you throw a pair of dice simultaneously 50 times, each act of throwing the pair is an event, so there are 50 events.

Suppose, in the process of tossing coins, you assign "heads" a value of 1 and "tails" a value of 0. Then when you toss a coin and it comes up "heads," you can say that the outcome of that event is 1. If you throw a pair of dice and get a sum total of 7, then the outcome of that event is 7.

The outcome of an event depends on the nature of the hardware and processes involved in the experiment. The use of a pair of "weighted" dice usually produces different outcomes, for an identical set of events, than a pair of "unweighted" dice. The outcome of an event also depends on how the event is defined. There is a difference between saying that the sum is 7 in a toss of two dice, as compared with saying that one of the die comes up 2 while the other one comes up 5.

SAMPLE SPACE

A *sample space* is the set of all possible outcomes in the course of an experiment. Even if the number of events is small, a sample space can be large.

Table 8-1 The sample space for an experiment in which a coin is tossed four times. There are 16 possible outcomes; "heads" = 1 and "tails" = 0.

Event 1	Event 2	Event 3	Event 4
0	0	0	0
0	0	0	1
0	0	1	0
0	0	1	1
0	1	0	0
0	1	0	1
0	1	1	0
0	1	1	1
1	0	0	0
1	0	0	1
1	0	1	0
1	0	1	1
1	1	0	0
1	1	0	1
1	1	1	0
1	1	1	1

If you toss a coin four times, there are 16 possible outcomes. These are listed in Table 8-1, where "heads" = 1 and "tails" = 0. (If the coin happens to land on its edge, you disregard that result and toss it again.)

If a pair of dice, one red and one blue, is tossed once, there are 36 possible outcomes in the sample space, as shown in Table 8-2. The outcomes are

Table 8-2 The sample space for an experiment consisting of a single
event, in which a pair of dice (one red, one blue) is tossed once. There
are 36 possible outcomes, shown as ordered pairs (red, blue).

Red → Blue ↓	1	2	3	4	5	6
1	(1,1)	(2,1)	(3,1)	(4,1)	(5,1)	(6,1)
2	(1,2)	(2,2)	(3,2)	(4,2)	(5,2)	(6,2)
3	(1,3)	(2,3)	(3,3)	(4,3)	(5,3)	(6,3)
4	(1,4)	(2,4)	(3,4)	(4,4)	(5,4)	(6,4)
5	(1,5)	(2,5)	(3,5)	(4,5)	(5,5)	(6,5)
6	(1,6)	(2,6)	(3,6)	(4,6)	(5,6)	(6,6)

denoted as ordered pairs, with the face-number of the red die listed first and
the face-number of the blue die listed second.

MATHEMATICAL PROBABILITY

Let x be a discrete random variable that can attain n possible values, all
equally likely. Suppose an outcome H results from exactly m different values
of x, where $m \leq n$. Then the *mathematical probability* $p_{math}(H)$ that outcome
H will result from any given value of x is given by the following formula:

$$p_{math}(H) = m/n$$

Expressed as a percentage, the probability $p_{\%}(H)$ is:

$$p_{math\%}(H) = 100m/n$$

If we toss an "unweighted" die once, each of the six faces is as likely to
turn up as each of the others. That is, we are as likely to see 1 as we are to
see 2, 3, 4, 5, or 6. In this case, there are 6 possible values, so $n=6$. The
mathematical probability of any one of the faces turning up ($m=1$) is equal
to $p_{math}(H)=1/6$. To calculate the mathematical probability of either of
any two different faces turning up (say 3 or 5), we set $m=2$; therefore
$p_{math}(H)=2/6=1/3$. If we want to know the mathematical probability that
any one of the six faces will turn up, we set $m=6$, so the formula gives us

$p_{math}(H) = 6/6 = 1$. The respective percentages $p_{math\%}(H)$ in these cases are 16.67% (approximately), 33.33% (approximately), and 100% (exactly).

Mathematical probabilities can only exist within the range 0 to 1 (or 0% to 100%) inclusive. The following formulas describe this constraint:

$$0 \leq p_{math}(H) \leq 1$$

$$0\% \leq p_{math\%}(H) \leq 100\%$$

We can never have a mathematical probability of 2, or −45%, or −6, or 556%. When you give this some thought, it is obvious. There is no way for something to happen less often than never. It's also impossible for something to happen more often than all the time.

EMPIRICAL PROBABILITY

In order to determine the likelihood that an event will have a certain outcome in real life, we must rely on the results of prior experiments. The chance of something happening based on experience or observation is called *empirical probability*.

Suppose we are told that a die is "unweighted." How does the person who tells us this know that it is true? If we want to use this die in some application, such as when we need an object that can help us to generate a string of random numbers from the set {1, 2, 3, 4, 5, 6}, we can't take on faith the notion that the die is "unweighted." We have to check it out. We can analyze the die in a lab and figure out where its center of gravity is; we measure how deep the indentations are where the dots on its faces are inked. We can scan the die electronically, X-ray it, and submerge it in (or float it on) water. But to be absolutely certain that the die is "unweighted," we must toss it many thousands of times, and be sure that each face turns up, on the average, 1/6 of the time. We must conduct an experiment – gather *empirical evidence* – that supports the contention that the die is "unweighted." Empirical probability is based on determinations of relative frequency, which was discussed in the last chapter.

As with mathematical probability, there are limits to the range an empirical probability figure can attain. If H is an outcome for a particular single event, and the empirical probability of H taking place as a result of that event is denoted $p_{emp}(H)$, then:

$$0 \leq p_{emp}(H) \leq 1$$

$$0\% \leq p_{emp\%}(H) \leq 100\%$$

PROBLEM 8-1

Suppose a new cholesterol-lowering drug comes on the market. If the drug is to be approved by the government for public use, it must be shown effective, and it must also be shown not to have too many serious side effects. So it is tested. During the course of testing, 10,000 people, all of whom have been diagnosed with high cholesterol, are given this drug. Imagine that 7289 of the people experience a significant drop in cholesterol. Also suppose that 307 of these people experience adverse side effects. If you have high cholesterol and go on this drug, what is the empirical probability $p_{emp}(B)$ that you will derive benefit? What is the empirical probability $p_{emp}(A)$ that you will experience adverse side effects?

SOLUTION 8-1

Some readers will say that this question cannot be satisfactorily answered because the experiment is not good enough. Is 10,000 test subjects a large enough number? What physiological factors affect the way the drug works? How about blood type, for example? Ethnicity? Gender? Blood pressure? Diet? What constitutes "high cholesterol"? What constitutes a "significant drop" in cholesterol level? What is an "adverse side effect"? What is the standard drug dose? How long must the drug be taken in order to know if it works? For convenience, we ignore all of these factors here, even though, in a true scientific experiment, it would be an excellent idea to take them all into consideration.

Based on the above experimental data, shallow as it is, the relative frequency of effectiveness is $7289/10,000 = 0.7289 = 72.89\%$. The relative frequency of ill effects is $307/10,000 = 0.0307 = 3.07\%$. We can round these off to 73% and 3%. These are the empirical probabilities that you will derive benefit, or experience adverse effects, if you take this drug in the hope of lowering your high cholesterol. Of course, once you actually use the drug, these probabilities will lose all their meaning for you. You will eventually say "The drug worked for me" or "The drug did not work for me." You will say, "I had bad side effects" or "I did not have bad side effects."

REAL-WORLD EMPIRICISM

Empirical probability is used by scientists to make predictions. It is not good for looking at aspects of the past or present. If you try to calculate the empirical probability of the existence of extraterrestrial life in our galaxy, you can play around with formulas based on expert opinions, but once you state a numeric figure, you commit the PF. If you say the empirical probability that a hurricane of category 3 or stronger struck the US mainland in 1992 equals x% (where $x < 100$) because at least one hurricane of that intensity hit the US mainland in x of the years in the 20th century, historians will tell you that is rubbish, as will anyone who was in Homestead, Florida on August 24, 1992.

Imperfection is inevitable in the real world. We can't observe an infinite number of people and take into account every possible factor in a drug test. We cannot toss a die an infinite number of times. The best we can hope for is an empirical probability figure that gets closer and closer to the "absolute truth" as we conduct a better and better experiment. Nothing we can conclude about the future is a "totally sure bet."

Properties of Outcomes

Here are some formulas that describe properties of outcomes in various types of situations. Don't let the symbology intimidate you. It is all based on the set theory notation covered in Chapter 1.

LAW OF LARGE NUMBERS

Suppose you toss an "unweighted" die many times. You get numbers turning up, apparently at random, from the set $\{1, 2, 3, 4, 5, 6\}$. What will the average value be? For example, if you toss the die 100 times, total up the numbers on the faces, and then divide by 100, what will you get? Call this number d (for die). It is reasonable to suppose that d will be fairly close to the mean, μ:

$$d \approx \mu$$
$$d \approx (1 + 2 + 3 + 4 + 5 + 6)/6$$
$$= 21/6$$
$$= 3.5$$

It's possible, in fact likely, that if you toss a die 100 times you'll get a value of d that is slightly more or less than 3.5. This is to be expected because of "reality imperfection." But now imagine tossing the die 1000 times, or 100,000 times, or even 100,000,000 times! The "reality imperfections" will be smoothed out by the fact that the number of tosses is so huge. The value of d will converge to 3.5. As the number of tosses increases without limit, the value of d will get closer and closer to 3.5, because the opportunity for repeated coincidences biasing the result will get smaller and smaller.

The foregoing scenario is an example of the *law of large numbers*. In a general, informal way, it can be stated like this: "As the number of events in an experiment increases, the average value of the outcome approaches the mean." This is one of the most important laws in all of probability theory.

INDEPENDENT OUTCOMES

Two outcomes H_1 and H_2 are *independent* if and only if the occurrence of one does not affect the probability that the other will occur. We write it this way:

$$p(H_1 \cap H_2) = p(H_1)p(H_2)$$

Figure 8-1 illustrates this situation in the form of a Venn diagram. The intersection is shown by the darkly shaded region.

A good example of independent outcomes is the tossing of a penny and a nickel. The face ("heads" or "tails") that turns up on the penny has no effect on the face ("heads" or "tails") that turns up on the nickel. It does not matter whether the two coins are tossed at the same time or at different times. They never interact with each other.

To illustrate how the above formula works in this situation, let $p(P)$ represent the probability that the penny turns up "heads" when a penny and a nickel are both tossed once. Clearly, $p(P) = 0.5$ (1 in 2). Let $p(N)$ represent the probability that the nickel turns up "heads" in the same scenario. It's obvious that $p(N) = 0.5$ (also 1 in 2). The probability that both coins turn up "heads" is, as you should be able to guess, 1 in 4, or 0.25. The above formula states it this way, where the intersection symbol $\cap$ can be translated as "and":

$$p(P \cap N) = p(P)p(N)$$
$$= 0.5 \times 0.5$$
$$= 0.25$$

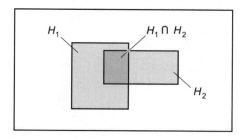

Fig. 8-1.   Venn diagram showing intersection.

MUTUALLY EXCLUSIVE OUTCOMES

Let H_1 and H_2 be two outcomes that are *mutually exclusive*; that is, they have no elements in common:

$$H_1 \cap H_2 = \varnothing$$

In this type of situation, the probability of either outcome occurring is equal to the sum of their individual probabilities. Here's how we write it, with the union symbol $\cup$ translated as "either/or":

$$p(H_1 \cup H_2) = p(H_1) + p(H_2)$$

Figure 8-2 shows this as a Venn diagram.

When two outcomes are mutually exclusive, they cannot both occur. A good example is the tossing of a single coin. It's impossible for "heads" and "tails" to both turn up on a given toss. But the sum of the two probabilities (0.5 for "heads" and 0.5 for "tails" if the coin is "balanced") is equal to the probability (1) that one or the other outcome will take place.

Another example is the result of a properly run, uncomplicated election for a political office between two candidates. Let's call the candidates Mrs. Anderson and Mr. Boyd. If Mrs. Anderson wins, we get outcome A, and if Mr. Boyd wins, we get outcome B. Let's call the respective probabilities of their winning $p(A)$ and $p(B)$. We might argue about the actual values of $p(A)$ and $p(B)$. We might obtain empirical probability figures by conducting a poll prior to the election, and get the idea that $p_{emp}(A) = 0.29$ and $p_{emp}(B) = 0.71$. The probability that either Mrs. Anderson or Mr. Boyd will win is equal to the sum of $p(A)$ and $p(B)$, whatever these values happen to be, and we can be sure it is equal to 1 (assuming neither of the candidates quits during the election and is replaced by a third, unknown person, and assuming there

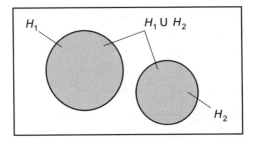

Fig. 8-2.   Venn diagram showing a pair of mutually exclusive outcomes.

are no write-ins or other election irregularities). Mathematically:

$$p(A \cup B) = p(A) + p(B)$$
$$= p_{\mathrm{emp}}(A) + P_{\mathrm{emp}}(B)$$
$$= 0.29 + 0.71$$
$$= 1$$

COMPLEMENTARY OUTCOMES

If two outcomes H_1 and H_2 are *complementary*, then the probability, expressed as a ratio, of one outcome is equal to 1 minus the probability, expressed as a ratio, of the other outcome. The following equations hold:

$$p(H_2) = 1 - p(H_1)$$
$$p(H_1) = 1 - p(H_2)$$

Expressed as percentages:

$$p_{\%}(H_2) = 100 - p_{\%}(H_1)$$
$$p_{\%}(H_1) = 100 - p_{\%}(H_2)$$

Figure 8-3 shows this as a Venn diagram.

The notion of complementary outcomes is useful when we want to find the probability that an outcome will fail to occur. Consider again the election between Mrs. Anderson and Mr. Boyd. Imagine that you are one of those peculiar voters who call themselves "contrarians," and who vote against, rather than for, candidates in elections. You are interested in the probability

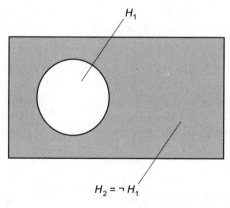

Fig. 8-3.   Venn diagram showing a pair of complementary outcomes.

that "your candidate" (the one you dislike more) will lose. According to the pre-election poll, $p_{emp}(A) = 0.29$ and $p_{emp}(B) = 0.71$. We might state this inside-out as:

$$p_{emp}(\neg B) = 1 - p_{emp}(B)$$
$$= 1 - 0.71$$
$$= 0.29$$

$$p_{emp}(\neg A) = 1 - p_{emp}(A)$$
$$= 1 - 0.29$$
$$= 0.71$$

where the "droopy minus sign" ($\neg$) stands for the "not" operation, also called *logical negation*. If you are fervently wishing for Mr. Boyd to lose, then you can guess from the poll that the likelihood of your being happy after the election is equal to $p_{emp}(\neg B)$, which is 0.29 in this case.

Note that in order for two outcomes to be complementary, the sum of their probabilities must be equal to 1. This means that one or the other (but not both) of the two outcomes must take place; they are the only two possible outcomes in a scenario.

NONDISJOINT OUTCOMES

Outcomes H_1 and H_2 are called *nondisjoint* if and only if they have at least one element in common:

$$H_1 \cap H_2 \neq \varnothing$$

In this sort of case, the probability of either outcome is equal to the sum of the probabilities of their occurring separately, minus the probability of their occurring simultaneously. The equation looks like this:

$$p(H_1 \cup H_2) = p(H_1) + p(H_2) - p(H_1 \cap H_2)$$

Figure 8-4 shows this as a Venn diagram. The intersection of probabilities is subtracted in order to ensure that the elements common to both sets (represented by the lightly shaded region where the two sets overlap) are counted only once.

PROBLEM 8-2
Imagine that a certain high school has 1000 students. The new swimming and diving coach, during his first day on the job, is looking for team prospects.

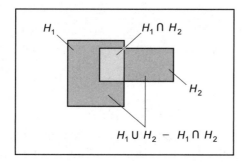

Fig. 8-4. Venn diagram showing a pair of nondisjoint outcomes.

Suppose that the following are true:

- 200 students can swim well enough to make the swimming team
- 100 students can dive well enough to make the diving team
- 30 students can make either team or both teams

If the coach wanders through the hallways blindfolded and picks a student at random, determine the probabilities, expressed as ratios, that the coach will pick

- a fast swimmer; call this $p(S)$
- a good diver; call this $p(D)$
- someone good at both swimming and diving; call this $p(S \cap D)$
- someone good at either swimming or diving, or both; call this $p(S \cup D)$

SOLUTION 8-2

This problem is a little tricky. We assume that the coach has objective criteria for evaluating prospective candidates for his teams! That having been said, we must note that the outcomes are not mutually exclusive, nor are they independent. There is overlap, and there is interaction. We can find the first three answers immediately, because we are told the numbers:

$$p(S) = 200/1000 = 0.200$$

$$p(D) = 100/1000 = 0.100$$

$$p(S \cap D) = 30/1000 = 0.030$$

In order to calculate the last answer – the total number of students who can make either team or both teams – we must find $p(S \cup D)$ using this formula:

$$p(S \cup D) = p(S) + p(D) - p(S \cap D)$$

$$= 0.200 + 0.100 - 0.030$$

$$= 0.270$$

This means that 270 of the students in the school are potential candidates for either or both teams. The answer is not 300, as one might at first expect. That would be the case only if there were no students good enough to make both teams. We mustn't count the exceptional students twice. (However well somebody can act like a porpoise, he or she is nevertheless only one person!)

MULTIPLE OUTCOMES

The formulas for determining the probabilities of mutually exclusive and nondisjoint outcomes can be extended to situations in which there are three possible outcomes.

Three mutually exclusive outcomes. Let H_1, H_2, and H_3 be three mutually exclusive outcomes, such that the following facts hold:

$$H_1 \cap H_2 = \varnothing$$

$$H_1 \cap H_3 = \varnothing$$

$$H_2 \cap H_3 = \varnothing$$

The probability of any one of the three outcomes occurring is equal to the sum of their individual probabilities (Fig. 8-5):

$$p(H_1 \cup H_2 \cup H_3) = p(H_1) + p(H_2) + p(H_3)$$

Three nondisjoint outcomes. Let H_1, H_2, and H_3 be three nondisjoint outcomes. This means that one or more of the following facts is true:

$$H_1 \cap H_2 \neq \varnothing$$

$$H_1 \cap H_3 \neq \varnothing$$

$$H_2 \cap H_3 \neq \varnothing$$

The probability of any one of the outcomes occurring is equal to the sum of the probabilities of their occurring separately, minus the probabilities of each

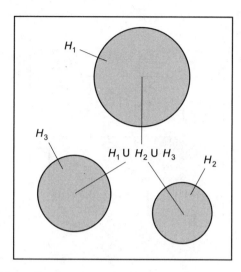

Fig. 8-5.   Venn diagram showing three mutually exclusive outcomes.

pair occurring simultaneously, minus the probability of all three occurring simultaneously (Fig. 8-6):

$$p(H_1 \cup H_2 \cup H_3)$$
$$= p(H_1) + p(H_2) + p(H_3)$$
$$\quad - p(H_1 \cap H_2) - p(H_1 \cap H_3) - p(H_2 \cap H_3)$$
$$\quad - p(H_1 \cap H_2 \cap H_3)$$

PROBLEM 8-3

Consider again the high school with 1000 students. The coach seeks people for the swimming, diving, and water polo teams in the same wandering, blindfolded way as before. Suppose the following is true of the students in the school:

- 200 people can make the swimming team
- 100 people can make the diving team
- 150 people can make the water polo team
- 30 people can make both the swimming and diving teams
- 110 people can make both the swimming and water polo teams
- 20 people can make both the diving and water polo teams
- 10 people can make all three teams

If the coach staggers around and tags students at random, what is the probability, expressed as a ratio, that the coach will, on any one tag, select a student who is good enough for at least one of the sports?

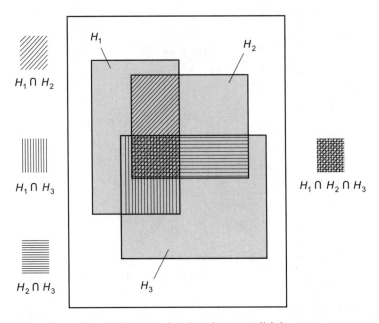

Fig. 8-6.   Venn diagram showing three nondisjoint outcomes.

SOLUTION 8-3

Let the following expressions stand for the respective probabilities, all representing the results of random selections by the coach (and all of which we are told):

- Probability that a student can swim fast enough $= p(S) = 200/1000 = 0.200$
- Probability that a student can dive well enough $= p(D) = 100/1000 = 0.100$
- Probability that a student can play water polo well enough $= p(W) = 150/1000 = 0.150$
- Probability that a student can swim fast enough and dive well enough $= p(S \cap D) = 30/1000 = 0.030$
- Probability that a student can swim fast enough and play water polo well enough $= p(S \cap W) = 110/1000 = 0.110$
- Probability that a student can dive well enough and play water polo well enough $= p(D \cap W) = 20/1000 = 0.020$
- Probability that a student can swim fast enough, dive well enough, and play water polo well enough $= p(S \cap D \cap W) = 10/1000 = 0.010$

In order to calculate the total number of students who can end up playing at least one sport for this coach, we must find $p(S \cup D \cup W)$ using this formula:

$$p(S \cup D \cup W) = p(S) + p(D) + p(W)$$
$$- p(S \cap D) - p(S \cap W) - p(D \cap W)$$
$$- p(S \cap D \cap W)$$
$$= 0.200 + 0.100 + 0.150$$
$$- 0.030 - 0.110 - 0.020 - 0.010$$
$$= 0.280$$

This means that 280 of the students in the school are potential prospects.

Permutations and Combinations

In probability, it is often necessary to choose small sets from large ones, or to figure out the number of ways in which certain sets of outcomes can take place. Permutations and combinations are the two most common ways this is done.

FACTORIAL

When working with multiple possibilities, it's necessary to be familiar with a function called the *factorial*. This function applies only to the natural numbers. (It can be extended to more values, but then it is called the *gamma function*.) The factorial of a number is indicated by writing an exclamation point after it.

If n is a natural number and $n \geq 1$, the value of $n!$ is defined as the product of all natural numbers less than or equal to n:

$$n! = 1 \times 2 \times 3 \times 4 \times \ldots \times n$$

If $n = 0$, then by convention, $n! = 1$. The factorial is not defined for negative numbers.

It's easy to see that as n increases, the value of $n!$ goes up rapidly, and when n reaches significant values, the factorial skyrockets. There is a formula for approximating $n!$ when n is large:

$$n! \approx n^n/e^n$$

where e is a constant called the *natural logarithm base*, and is equal to approximately 2.71828. The squiggly equals sign emphasizes the fact that the value of $n!$ using this formula is approximate, not exact.

PROBLEM 8-4

Write down the values of the factorial function for $n = 0$ through $n = 15$, in order to illustrate just how fast this value "blows up."

SOLUTION 8-4

The results are shown in Table 8-3. It's perfectly all right to use a calculator here. It should be capable of displaying a lot of digits. Most personal computers have calculators that are good enough for this purpose.

Table 8-3   Values of $n!$ for $n = 0$ through $n = 15$. This
table constitutes the solution to Problem 8-4.

Value of n	Value of $n!$
0	0
1	1
2	2
3	6
4	24
5	120
6	720
7	5040
8	40,320
9	362,880
10	3,628,800
11	39,916,800
12	479,001,600
13	6,227,020,800
14	87,178,291,200
15	1,307,674,368,000

PROBLEM 8-5

Determine the approximate value of 100! using the formula given above.

SOLUTION 8-5

A calculator is not an option here; it is a requirement. You should use one that has an e^x (or natural exponential) function key. In case your calculator does not have this key, the value of the exponential function can be found by using the natural logarithm key and the inverse function key together. It will also be necessary for the calculator to have an x^y key (also called $x\char`^y$) that lets you find the value of a number raised to its own power. In addition, the calculator should be capable of displaying numbers in *scientific notation*, also called *power-of-10 notation*. Most personal computer calculators are adequate if they are set for scientific mode.

Using the above formula for $n = 100$:

$$100! \approx (100^{100})/e^{100}$$

$$\approx (1.00 \times 10^{200})/(2.688117 \times 10^{43})$$

$$\approx 3.72 \times 10^{156}$$

The numeral representing this number, if written out in full, would be a string of digits too long to fit on most text pages without taking up two or more lines. Your calculator will probably display it as something like 3.72e+156 or 3.72 E 156. In these displays, the "e" or "E" does not refer to the natural logarithm base. Instead, it means "times 10 raised to the power of."

PERMUTATIONS

When working with problems in which items are taken from a larger set in specific order, the idea of a permutation is useful. Suppose q and r are both positive integers. Let q represent a set of items or objects taken r at a time in a specific order. The possible number of permutations in this situation is symbolized $_qP_r$ and can be calculated as follows:

$$_qP_r = q!/(q-r)!$$

COMBINATIONS

Let q represent a set of items or objects taken r at a time in no particular order, and where both q and r are positive integers. The possible number of combinations in this situation is symbolized $_qC_r$ and can be calculated as follows:

$$_qC_r = {_qP_r}/r! = q!/[r!(q-r)!]$$

PROBLEM 8-6

How many permutations are there if you have 10 apples, taken 5 at a time in a specific order?

SOLUTION 8-6

Use the above formula for permutations, plugging in $q=10$ and $r=5$:

$$_{10}P_5 = 10!/(10-5)!$$
$$= 10!/5!$$
$$= 10 \times 9 \times 8 \times 7 \times 6$$
$$= 30,240$$

PROBLEM 8-7

How many combinations are there if you have 10 apples, taken 5 at a time in no particular order?

SOLUTION 8-7

Use the above formula for combinations, plugging in $q=10$ and $r=5$. We can use the formula that derives combinations based on permutations, because we already know from the previous problem that $_{10}P_5 = 30,240$:

$$_{10}C_5 = {_{10}P_5}/5!$$
$$= 30,240/120$$
$$= 252$$

Quiz

Refer to the text in this chapter if necessary. A good score is eight correct. Answers are in the back of the book.

1. Empirical probability is based on
   (a) observation or experimentation
   (b) theoretical models only

(c) continuous outcomes

(d) standard deviations

2. What is the number of possible combinations of 7 objects taken 3 at a time?

    (a) 10

    (b) 21

    (c) 35

    (d) 210

3. What is the number of possible permutations of 7 objects taken 3 at a time?

    (a) 10

    (b) 21

    (c) 35

    (d) 210

4. The difference between permutations and combinations lies in the fact that

    (a) permutations take order into account, but combinations do not

    (b) combinations take order into account, but permutations do not

    (c) combinations involve only continuous variables, but permutations involve only discrete variables

    (d) permutations involve only continuous variables, but combinations involve only discrete variables

5. The result of an event is called

    (a) an experiment

    (b) a trial

    (c) an outcome

    (d) a variable

6. How many times as large is 1,000,000 factorial, compared with 999,999 factorial?

    (a) 1,000,000 times as large.

    (b) 999,999 times as large.

    (c) A huge number that requires either a computer or else many human-hours to calculate.

    (d) There is not enough information given here to get any idea.

7. The set of all possible outcomes during the course of an experiment is called

    (a) a dependent variable

    (b) a random variable

(c) a discrete variable

(d) a sample space

8. What is the mathematical probability that a coin, tossed 10 times in a row, will come up "tails" on all 10 tosses?

   (a) 1/10

   (b) 1/64

   (c) 1/1024

   (d) 1/2048

9. Two outcomes are mutually exclusive if and only if

   (a) they are nondisjoint

   (b) they have no elements in common

   (c) they have at least one element in common

   (d) they have identical sets of outcomes

10. The probability, expressed as a percentage, of a particular occurrence can never be

   (a) less than 100

   (b) less than 0

   (c) greater than 1

   (d) anything but a whole number

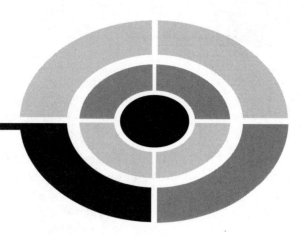

Test: Part 2

Do not refer to the text when taking this test. You may draw diagrams or use a calculator if necessary. A good score is at least 30 correct. Answers are in the back of the book. It's best to have a friend check your score the first time, so you won't memorize the answers if you want to take the test again.

1. Consider the following pair of equations, in which the intent is to find a solution or solutions, if any, for x and y simultaneously:

    $$y = 3x + 4$$
    $$y = 3x - 2$$

    How many real-number solutions are there in this case?
    (a) None.
    (b) One.
    (c) More than one.
    (d) Infinitely many.
    (e) More information is needed in order to answer this question.

2. In the complex number $2 + j3$, the imaginary component consists of
    (a) the number 2

(b) the number 3

(c) the number 2 multiplied by the square root of −1

(d) the number 3 multiplied by the square root of −1

(e) the product of 2 and 3, which is 6

3. In a frequency distribution, frequency is portrayed as the
    (a) cumulative variable
    (b) discrete variable
    (c) dependent variable
    (d) random variable
    (e) continuous variable

4. What is the mathematical probability that a coin, tossed five times in
    a row, will come up "heads" on all five tosses?
    (a) 1
    (b) 1/2
    (c) 1/5
    (d) 1/25
    (e) 1/32

5. Fill in the blank to make this sentence true: "A sample of a population
    is _____ of that population."
    (a) an expansion
    (b) a subset
    (c) a variance
    (d) the average
    (e) the median

6. In Fig. Test 2-1, suppose H_1 and H_2 represent two different sets of out-
    comes in an experiment. The light-shaded region, labeled H_1, represents
    (a) the set of outcomes belonging to H_1 but not to H_2
    (b) the set of outcomes belonging to neither H_1 nor H_2

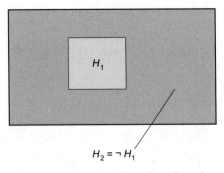

H_1

$H_2 = \neg H_1$

Fig. Test 2-1. Illustration for Part Two Test Questions 6 through 9.

(c) the set of outcomes belonging to either H_1 or H_2, but not both
(d) the set of outcomes belonging to either H_1 or H_2, or both
(e) the set of outcomes belonging to H_2 but not to H_1

7. In Fig. Test 2-1, the dark-shaded region, labeled $H_2 = \neg H_1$, represents
    (a) the set of outcomes common to both H_1 and H_2
    (b) the set of outcomes belonging to neither H_1 nor H_2
    (c) the set of outcomes belonging to either H_1 or H_2, but not both
    (d) the set of outcomes belonging to either H_1 or H_2, or both
    (e) the set of outcomes belonging to H_2 but not to H_1

8. In Fig. Test 2-1, the entire portion that is shaded, either light or dark, represents
    (a) the set of outcomes common to both H_1 and H_2
    (b) the set of outcomes belonging to neither H_1 nor H_2
    (c) the set of outcomes belonging to either H_1 or H_2, but not both
    (d) the empty set
    (e) the set of outcomes belonging to H_2 but not to H_1

9. The Venn diagram of Fig. Test 2-1 portrays
    (a) complementary outcomes
    (b) mutually exclusive outcomes
    (c) independent outcomes
    (d) coincident outcomes
    (e) nondisjoint outcomes

10. Which of the following is not an allowable maneuver to perform in solving a single-variable equation?
    (a) Adding a constant to each side.
    (b) Subtracting a constant from each side.
    (c) Dividing each side by a variable that may attain a value of 0.
    (d) Multiplying each side by a variable that may attain a value of 0.
    (e) All of the above (a), (b), (c), and (d) are allowable.

11. What is the solution of the equation $3x + 5 = 2x - 4$?
    (a) There isn't any.
    (b) More information is needed in order to find out.
    (c) 9
    (d) -9
    (e) 9 or -9

12. What is the mathematical probability that an "unweighted" die, tossed three times, will show the face with 2 dots on all three occasions?
    (a) 1 in 6

(b) 1 in 36
(c) 1 in 64
(d) 1 in 216
(e) 1 in 46,656

13. What is the solution set of the equation $x^2 + 64 = 0$?
(a) The empty set.
(b) $\{8\}$
(c) $\{8, -8\}$
(d) $\{j8, -j8\}$
(e) More information is needed to answer this.

14. Fill in the blank to make the following sentence true: "If the number of elements in a distribution is odd, then the _____ is the value such that the number of elements having values greater than or equal to it is the same as the number of elements having values less than or equal to it."
(a) mean
(b) deviation
(c) frequency
(d) mode
(e) median

15. Let q represent a set of items or objects taken r at a time in a specific order. The possible number of permutations in this situation is symbolized $_qP_r$ and can be calculated as follows:

$$_qP_r = q!/(q - r)!$$

Given this information, what is the possible number of permutations of 8 objects taken 2 at a time?
(a) 40,320
(b) 720
(c) 56
(d) 28
(e) 24

16. Let q represent a set of items or objects taken r at a time in no particular order, where $q \leq r$, and where both q and r are positive integers. The possible number of combinations in this situation is symbolized $_qC_r$ and can be calculated as follows:

$$_qC_r = q!/[r!(q - r)!]$$

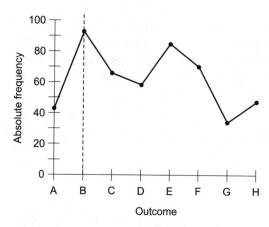

Fig. Test 2-2.   Illustration for Part Two Test Questions 17 through 19.

Given this information, what is the possible number of combinations of 8 objects taken 2 at a time?
(a) 40,320
(b) 720
(c) 56
(d) 28
(e) 24

17.   Fig. Test 2-2 shows the results of a hypothetical experiment with eight possible outcomes, called A through H. This illustration is an example of
(a) a vertical bar graph
(b) a point-to-point graph
(c) a horizontal bar graph
(d) a histogram
(e) linear interpolation

18.   In Fig. Test 2-2, the vertical, dashed line passes through a point corresponding to
(a) the mean
(b) the median
(c) the mode
(d) the distribution factor
(e) the deviation

19.   What, if anything, is wrong with the graph of Fig. Test 2-2?
(a) If the frequency values of the points are added up, the sum should be equal to 100, but it obviously isn't.

(b) A set of events with outcomes like this cannot possibly occur.

(c) The highest frequency value should be in the middle, and the values to either side should decrease steadily.

(d) The frequency values should increase steadily as you move toward the right in the graph.

(e) There is nothing wrong with this graph.

20. What is the solution to the equation $3x + 4y = 12$?
    (a) 3
    (b) 4
    (c) −3
    (d) −4
    (e) There is no unique solution to this equation.

21. The equation $3x^2 + 5x + 3 = 0$ is an example of
    (a) a linear equation
    (b) a quadratic equation
    (c) a cubic equation
    (d) a quartic equation
    (e) a multivariable equation

22. In statistics, a discrete variable
    (a) can attain specific, defined values
    (b) can attain infinitely many values within a defined span
    (c) can attain only positive-integer values
    (d) can attain only non-negative integer values
    (e) can never have a value greater than 1

23. Fill in the blank to make the following sentence true: "In a statistical distribution having discrete values for the independent variable, the highest cumulative absolute frequency number should be equal to the _____ of all the individual absolute frequency numbers."
    (a) average
    (b) sum
    (c) factorial
    (d) product
    (e) largest

24. In the equation $4x^3 + 8x^2 - 17x - 6 = 0$, the numbers 4, 8, −17, and −6 are called the
    (a) factors
    (b) solutions
    (c) coefficients

(d) complex numbers
(e) imaginary numbers

25. If two outcomes H_1 and H_2 are complementary, then the probability, expressed as a percentage, of one outcome is equal to
(a) the probability of the other outcome
(b) 100% plus the probability of the other outcome
(c) 100% minus the probability of the other outcome
(d) 100% times the probability of the other outcome
(e) 100% divided by the probability of the other outcome

26. What can be said about the equation $(x-3)(x+2)(x+6)=0$?
(a) It is a linear equation.
(b) It is a quadratic equation.
(c) It is a cubic equation.
(d) It is a quartic equation.
(e) None of the above.

27. Consider the equation $(x-3)(x+2)(x+6)=0$. By examining this, it is apparent that
(a) there are no real-number solutions
(b) there exists exactly one real-number solution
(c) there exist exactly two different real-number solutions
(d) there exist exactly three different real-number solutions
(e) there exist infinitely many different real-number solutions

28. Suppose the two lines graphed in Fig. Test 2-3 represent linear equations. It appears that for the equation of line A, when $y=0$, the value of x
(a) is not defined

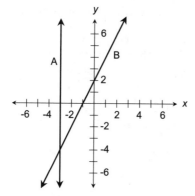

Fig. Test 2-3. Illustration for Part Two Test Questions 28 through 31.

(b) is positive

(c) is negative

(d) can be any real number

(e) requires more information to be defined

29. Suppose the two lines graphed in Fig. Test 2-3 represent linear equations. It appears that for the equation of line A, when y is positive, the value of x

(a) is not defined

(b) is positive

(c) is negative

(d) can be any real number

(e) requires more information to be defined

30. Suppose the two lines graphed in Fig. Test 2-3 represent linear equations. It appears that for the equation of line B, when $x > -1$, the value of y

(a) is not defined

(b) is positive

(c) is negative

(d) can be any real number

(e) requires more information to be defined

31. Suppose the two lines graphed in Fig. Test 2-3 represent linear equations. It appears that the two equations, considered simultaneously:

(a) have no solutions

(b) have a single solution

(c) have two solutions

(d) have infinitely many solutions

(e) requires more information to be defined

32. The term *fuzzy truth* is used to describe

(a) a theory in which there are degrees of truth that span a range

(b) standard deviation

(c) cumulative frequency

(d) the probability fallacy

(e) any continuous distribution

33. What is the solution set of the equation $x^2 - 12x = -35$?

(a) $\{12, 35\}$

(b) $\{-12, -35\}$

(c) $\{1, -12, -35\}$

(d) $\{-1, 12, 35\}$

(e) None of the above.

34. The mean for a discrete variable in a distribution is
    (a) the mathematical average of the values
    (b) the middle value
    (c) half of the sum of the values
    (d) the square root of the product of all the values
    (e) the reciprocal of the sum of the values

35. In statistics, the frequency of an outcome is
    (a) the number of cycles per second during which the outcome occurs
    (b) the likelihood that an outcome will occur more than once
    (c) an outcome chosen at random
    (d) the number of times the outcome occurs within a sample
    (e) an undefined term

36. What is the solution set of the equation $(x - 4)^2 = 0$?
    (a) $\{4\}$
    (b) $\{-4\}$
    (c) $\{4, -4\}$
    (d) $\varnothing$
    (e) It is impossible to tell because this equation is not in standard or factored form.

37. Consider the following pair of equations, in which the intent is to find a solution or solutions, if any, for x and y simultaneously:

    $$3x - 4y = 8$$
    $$6x + 2y = -4$$

    This is an example of a set of
    (a) two-by-two linear equations
    (b) two-by-two quadratic equations
    (c) cubic equations
    (d) nonlinear equations
    (e) redundant equations

38. In Fig. Test 2-4, suppose that curves A and B represent equations for y versus x. How many solutions are there to both equations considered simultaneously?
    (a) None.
    (b) One.
    (c) More than one.

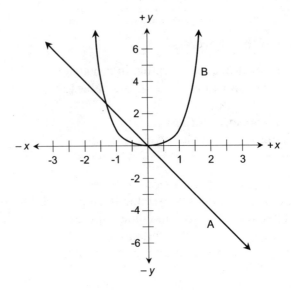

Fig. Test 2-4. Illustration for Part Two Test Questions 38 through 40.

(d) Infinitely many.

(e) More information is needed to answer this question.

39. In Fig. Test 2-4, suppose curve A represents the equation $x + y = 0$, while curve B represents the equation $x^4 - 2y = 0$. Which of the following ordered pairs, representing values (x,y), is a solution to both of these equations considered simultaneously?

   (a) $(0,0)$

   (b) $(-1,1)$

   (c) $(1,1)$

   (d) $(2,4)$

   (e) None of the above.

40. In Fig. Test 2-4, suppose that curves A and B represent equations for y versus x. What can be said in general about these equations?

   (a) Neither of them is linear.

   (b) One of them is linear.

   (c) Both of them are linear.

   (d) Both of them are quadratic.

   (e) Both of them are quartic.

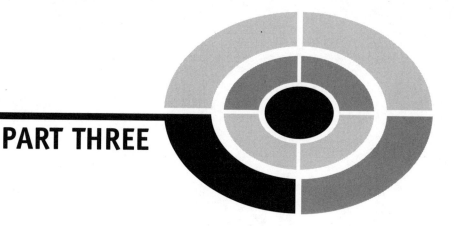

PART THREE

Shapes and Places

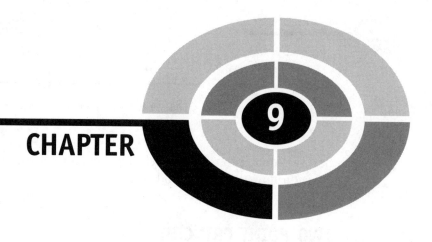

Geometry on the Flats

Geometry is widely used in science and engineering. The ancient Egyptians and Greeks used geometry to calculate the diameter of the earth and the distance to the moon, and scientists have been busy with it ever since. You do not have to memorize all the formulas in this chapter, but it's good to be comfortable with this sort of stuff. So roll up your sleeves and make sure your brain is working.

Fundamental Rules

Here are some of the most important principles of *plane geometry*, which involves the behavior of points, lines, and various figures confined to a

Fig. 9-1.   Two point principle.

flat, two-dimensional (2D) surface. There are lots of illustrations to help you envision how these rules and formulas work.

TWO POINT PRINCIPLE

Suppose that P and Q are two distinct geometric points. Then the following statements are true, as shown in Fig. 9-1:

- P and Q lie on a common straight line L.
- L is the only straight line on which both points lie.

THREE POINT PRINCIPLE

Let P, Q, and R be three distinct points, not all of which lie on a straight line. Then the following statements are true:

- P, Q, and R all lie in a common Euclidean (flat) plane S.
- S is the only Euclidean plane in which all three points lie.

PRINCIPLE OF n POINTS

Let P_1, P_2, $P_3, \ldots, P_n$ be n distinct points, not all of which lie in the same Euclidean (that is, "non-warped") space of $n-1$ dimensions. Then the following statements are true:

- P_1, P_2, $P_3, \ldots, P_n$ all lie in a common Euclidean space U of n dimensions.
- U is the only n-dimensional Euclidean space in which all n points lie.

DISTANCE NOTATION

The distance between any two distinct points P and Q, as measured from P towards Q along the straight line connecting them, is symbolized by writing PQ.

MIDPOINT PRINCIPLE

Suppose there is a straight line segment connecting two distinct points P and R. Then there is one and only one point Q on the line segment, between P and R, such that $PQ = QR$. This is illustrated in Fig. 9-2.

ANGLE NOTATION

Imagine that P, Q, and R are three distinct points. Let L be the straight line segment connecting P and Q; let M be the straight line segment connecting R and Q. Then the angle between L and M, as measured at point Q in the plane defined by the three points, can be written as $\angle PQR$ or as $\angle RQP$. If the rotational sense of measurement is specified, then $\angle PQR$ indicates the angle as measured from L to M, and $\angle RQP$ indicates the angle as measured from M to L (Fig. 9-3). These notations can also stand for the measures of angles, expressed either in degrees or in radians.

ANGLE BISECTION

Suppose there is an angle $\angle PQR$ measuring less than $180°$ and defined by three distinct points P, Q, and R, as shown in Fig. 9-4. Then there is exactly

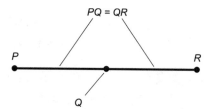

Fig. 9-2.   Midpoint principle.

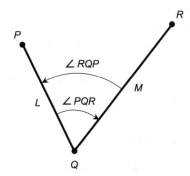

Fig. 9-3.   Angle notation and measurement.

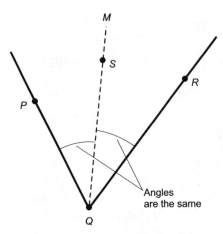

Fig. 9-4. Angle bisection principle.

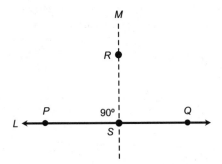

Fig. 9-5. Perpendicular principle.

one straight ray M that bisects the angle $\angle PQR$. If S is any point on M other than the point Q, then $\angle PQS = \angle SQR$. Every angle has one and only one ray that divides the angle in half.

PERPENDICULARITY

Suppose that L is a straight line through two distinct points P and Q. Let R be a point not on L. Then there is one and only one straight line M through point R, intersecting line L at some point S, such that M is perpendicular to L. This is shown in Fig. 9-5.

PERPENDICULAR BISECTOR

Suppose that L is a straight line segment connecting two distinct points P and R. Then there is one and only one straight line M that intersects line

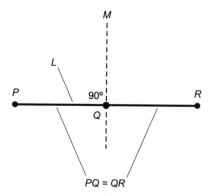

Fig. 9-6.   Perpendicular bisector principle.

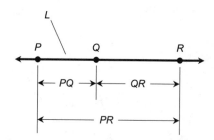

Fig. 9-7.   Distance addition and subtraction.

segment L in a point Q, such that M is perpendicular to L and the distance from P to Q is equal to the distance from Q to R. That is, every line segment has exactly one perpendicular bisector. This is illustrated in Fig. 9-6.

DISTANCE ADDITION AND SUBTRACTION

Let P, Q, and R be distinct points on a straight line L, such that Q is between P and R. Then the following equations hold concerning distances as measured along L (Fig. 9-7):

$$PQ + QR = PR$$
$$PR - PQ = QR$$
$$PR - QR = PQ$$

ANGLE ADDITION AND SUBTRACTION

Let P, Q, R, and S be four distinct points that lie in a common Euclidean plane. Let Q be the *vertex* of three angles $\angle PQR$, $\angle PQS$, and $\angle SQR$, as

shown in Fig. 9-8. Then the following equations hold concerning the angular measures:

$$\angle PQS + \angle SQR = \angle PQR$$
$$\angle PQR - \angle PQS = \angle SQR$$
$$\angle PQR - \angle SQR = \angle PQS$$

VERTICAL ANGLES

Suppose that L and M are two straight lines that intersect at a point P. Opposing pairs of angles, denoted x and y in Fig. 9-9, are known as *vertical angles* and always have equal measure.

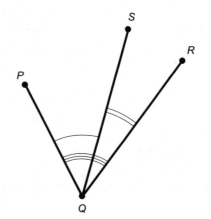

Fig. 9-8.   Angular addition and subtraction.

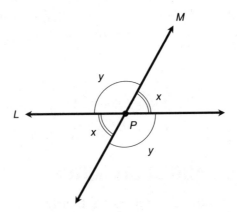

Fig. 9-9.   Vertical angles have equal measure.

ALTERNATE INTERIOR ANGLES

Suppose that L and M are parallel, straight lines. Let N be a straight *transversal line* that intersects L and M at points P and Q, respectively. In Fig. 9-10, angles labeled x are *alternate interior angles*; the same holds true for angles labeled y. Alternate interior angles always have equal measure. The transversal line N is perpendicular to lines L and M if and only if $x = y$.

ALTERNATE EXTERIOR ANGLES

Suppose that L and M are parallel, straight lines. Let N be a straight transversal line that intersects L and M at points P and Q, respectively. In Fig. 9-11, angles labeled x are *alternate exterior angles*; the same holds true for angles labeled y. Alternate exterior angles always have

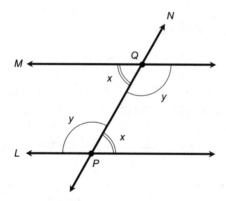

Fig. 9-10.   Alternate interior angles have equal measure.

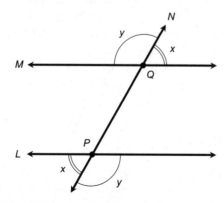

Fig. 9-11.   Alternate exterior angles have equal measure.

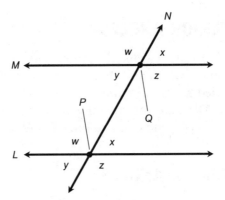

Fig. 9-12.   Corresponding angles have equal measure.

equal measure. The transversal line N is perpendicular to lines L and M if and only if $x = y$.

CORRESPONDING ANGLES

Let L and M be parallel, straight lines. Let N be a straight transversal line that intersects L and M at points P and Q, respectively. In Fig. 9-12, angles labeled w are *corresponding angles*; the same holds true for angles labeled x, y, and z. Corresponding angles always have equal measure. The transversal line N is perpendicular to lines L and M if and only if $w = x = y = z = 90°$; that is, if and only if all four angles are right angles.

PARALLEL PRINCIPLE

Suppose L is a straight line, and P is a point not on L. Then there exists one and only one straight line M through P, such that line M is parallel to line L (Fig. 9-13). This is one of the most important postulates in *Euclidean geometry*. Its negation can take two forms: either there is no such straight line M, or there exist two or more such straight lines M_1, M_2, M_3, ... Either form of the negation of this principle constitutes a cornerstone of *non-Euclidean geometry*, which is important to physicists and astronomers interested in the theories of general relativity and cosmology.

MUTUAL PERPENDICULARITY

Let L and M be straight lines that lie in the same plane. Suppose both L and M intersect a third straight line N, and both L and M are perpendicular to N. Then lines L and M are parallel to each other (Fig. 9-14).

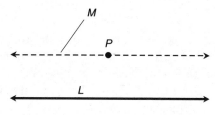

Fig. 9-13.   The parallel principle.

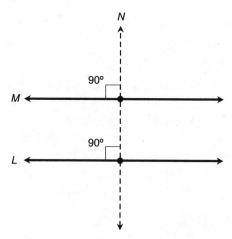

Fig. 9-14.   Mutual perpendicularity.

PROBLEM 9-1

Suppose a straight, flat, two-lane highway crosses a straight set of railroad tracks, such that the *acute angles* (angles measuring more than 0° but less than 90°) between the rails and the side lines of the highway measure 50°. What are the measures of the *obtuse angles* (angles measuring more than 90° but less than 180°) between the rails and the side lines of the highway?

SOLUTION 9-1

When a straight line is crossed by a straight transversal, the measures of adjacent angles at the intersection point always add up to 180°. The rails are straight, as are the side lines of the highway. Therefore, the obtuse angles measure 180° − 50°, or 130°. You may wish to draw a picture of this situation to make it easier to envision.

PROBLEM 9-2

In the situation of the previous problem, what are the measures of the acute angles between the highway center line and the rails? What are the measures of the obtuse angles between the highway center line and the rails?

SOLUTION 9-2

The highway center line is straight, and is parallel to both of the side lines. Thus, the acute angles between the rails and the center line have the same measures as the acute angles between the rails and the side lines, that is, 50°. The measures of the obtuse angles between the rails and the center line are the same as the measures of the obtuse angles between the rails and the side lines, that is, $180° - 50° = 130°$.

Triangles

If it's been a while since you took a course in plane geometry, you probably think of triangles when the subject is brought up. Maybe you recall having to learn all kinds of theoretical proofs concerning triangles, using "steps/ reasons" tables if your teacher was rigid, and less formal methods if your teacher was not so stodgy. You won't have to go through the proofs again here, but some of the more important facts about triangles are worth stating.

POINT–POINT–POINT

Let P, Q, and R be three distinct points, not all of which lie on the same straight line. Then the following statements are true (Fig. 9-15):

- P, Q, and R lie at the vertices of a triangle.
- This is the only triangle having vertices P, Q, and R.

SIDE–SIDE–SIDE

Let S, T, and U be three distinct, straight line segments. Let s, t, and u be the lengths of those three line segments, respectively. Suppose that S, T, and U

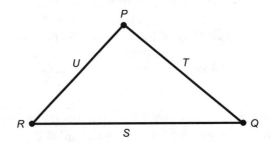

Fig. 9-15.   The three point principle; side–side–side triangles.

are joined at their end points P, Q, and R (Fig. 9-15). Then the following statements hold true:

- Line segments S, T, and U determine a triangle.
- This is the only triangle of its size and shape that has sides S, T, and U.
- All triangles having sides of lengths s, t, and u are *congruent* (identical in size and shape).

SIDE–ANGLE–SIDE

Let S and T be two distinct, straight line segments. Let P be a point that lies at the ends of both of these line segments. Denote the lengths of S and T by their lowercase counterparts s and t, respectively. Suppose S and T subtend an angle x relative to each other at point P (Fig. 9-16). Then the following statements are all true:

- S, T, and x determine a triangle.
- This is the only triangle having sides S and T that subtend an angle x at point P.
- All triangles containing two sides of lengths s and t that subtend an angle x are congruent.

ANGLE–SIDE–ANGLE

Let S be a straight line segment having length s, and whose end points are P and Q. Let x and y be the angles subtended relative to S by two straight lines L and M that run through P and Q, respectively (Fig. 9-17). Then the following statements are all true:

- S, x, and y determine a triangle.
- This is the only triangle determined by S, x, and y.

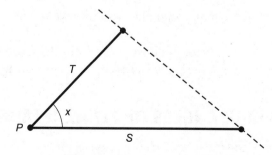

Fig. 9-16.   Side–angle–side triangles.

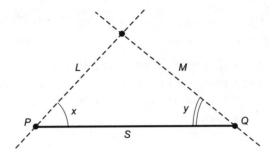

Fig. 9-17.   Angle–side–angle triangles.

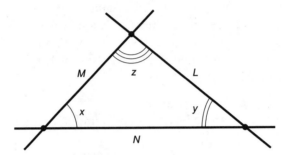

Fig. 9-18.   Angle–angle–angle triangles.

- All triangles containing one side of length s, and whose other two sides subtend angles of x and y relative to the side whose length is s, are congruent.

ANGLE–ANGLE–ANGLE

Let L, M, and N be straight lines that lie in a common plane and intersect in three distinct points, as illustrated in Fig. 9-18. Let the angles at these points be x, y, and z. Then the following statements are true:

- There are infinitely many triangles with interior angles x, y, and z in the sense shown.
- All triangles with interior angles x, y, and z in the sense shown are similar (that is, they have the same shape, but not necessarily the same size).

SUM OF INTERIOR ANGLES OF PLANE TRIANGLE

Suppose we have a general plane triangle, with no special properties other than the fact that it is defined by three distinct vertex points. Let the measures

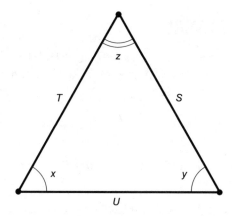

Fig. 9-19.   Isosceles and equilateral triangles.

of the interior angles be x, y, and z. The following equation holds if the angular measures are given in degrees:

$$x + y + z = 180°$$

ISOSCELES TRIANGLE

Suppose we have a triangle with sides S, T, and U, having lengths s, t, and u. Let x, y, and z be the angles opposite S, T, and U respectively (Fig. 9-19). Suppose any of the following equations hold:

$$s = t$$
$$t = u$$
$$s = u$$
$$x = y$$
$$y = z$$
$$x = z$$

Then the triangle is an *isosceles triangle*, and the following logical statements are true:

$$\text{If } s = t \text{ then } x = y$$
$$\text{If } t = u \text{ then } y = z$$
$$\text{If } s = u \text{ then } x = z$$
$$\text{If } x = y \text{ then } s = t$$
$$\text{If } y = z \text{ then } t = u$$
$$\text{If } x = z \text{ then } s = u$$

EQUILATERAL TRIANGLE

Suppose we have a triangle with sides S, T, and U, having lengths s, t, and u. Let x, y, and z be the angles opposite S, T, and U respectively (Fig. 9-19). Suppose either of the following are true:

$$s = t = u$$
$$x = y = z$$

Then the triangle is said to be an *equilateral triangle*, and the following logical statements are valid:

$$\text{If } s = t = u \text{ then } x = y = z$$
$$\text{If } x = y = z \text{ then } s = t = u$$

That is, all equilateral triangles have precisely the same shape; they are all similar. They are not necessarily all the same size, however.

THEOREM OF PYTHAGORAS

Suppose we have a right triangle defined by points P, Q, and R whose sides are D, E, and F having lengths d, e, and f, respectively. Let f be the side opposite the right angle (Fig. 9-20). Then the following equation is always true:

$$d^2 + e^2 = f^2$$

The converse of this is also true: If there is a triangle whose sides have lengths d, e, and f, and the above equation is true, then that triangle is a right triangle.

The longest side of a right triangle is always the side opposite the right angle, and is called the *hypotenuse*. The above formula can be stated verbally as follows: "The square of the length of the hypotenuse of a right triangle

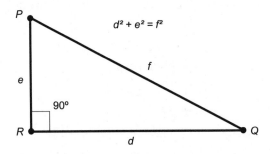

Fig. 9-20. The Theorem of Pythagoras.

is equal to the sum of the squares of the lengths of the other two sides." This is called the *Theorem of Pythagoras* or the *Pythagorean Theorem*. It is named after a Greek geometer, *Pythagoras of Samos*, who lived in the 4th century B.C.

PERIMETER OF TRIANGLE

Suppose we have a triangle defined by points P, Q, and R, and having sides S, T, and U of lengths s, t, and u, as shown in Fig. 9-21. Let s be the base length, h be the height, and x be the angle between the sides having length s and t. Then the perimeter, B, of the triangle is given by the following formula:

$$B = s + t + u$$

INTERIOR AREA OF TRIANGLE

Consider the same triangle as defined above; refer again to Fig. 9-21. The interior area, A, can be found with this formula:

$$A = sh/2$$

PROBLEM 9-3

What is the interior area of a triangle whose base is 4 meters (m) long and whose height is 2 m?

SOLUTION 9-3

In this case, the length s is 4 (in meters) and the height h is 2 (also in meters). Therefore, the area (in square meters or m^2) is:

$$A = sh/2$$
$$= 4 \times 2/2$$
$$= 4\,\text{m}^2$$

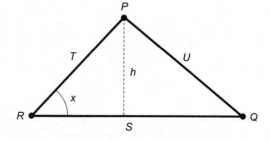

Fig. 9-21.　Perimeter and area of triangle.

PROBLEM 9-4

If the height and the base length of a triangle are both doubled, what happens to the interior area?

SOLUTION 9-4

The interior area quadruples. Let's prove this for the general case. Let the initial base length be s_1 and the initial height be h_1. Let the final base length, s_2, be $2s_1$ (twice the initial base length) and the final height, h_2, be $2h_1$ (twice the initial height):

$$s_2 = 2s_1$$
$$h_2 = 2h_1$$

Then the initial area, A_1, is:

$$A_1 = s_1 h_1/2$$

The final area, A_2, is:

$$A_2 = s_2 h_2/2$$
$$= (2s_1)(2h_1)/2$$
$$= 4(s_1 h_1/2)$$
$$= 4A_1$$

This shows that doubling both the base length and the height of a triangle causes its interior area to increase by a factor of 4. Interestingly, this is true even if the shape of the triangle changes in the process of the base-length and height change. As long as the constraints are met (doubling of both the base length and the height), it doesn't matter if, or how much, the triangle is "stretched horizontally"; the interior area will increase by the same factor regardless.

Quadrilaterals

A four-sided geometric figure that lies in a single plane is called a *quadrilateral*. There are several classifications, and various formulas that apply to each.

PARALLELOGRAM DIAGONALS

Suppose we have a parallelogram defined by four distinct points P, Q, R, and S. Let D be a straight line segment connecting P and R as shown in Fig. 9-22A. Then D is a *minor diagonal* of the parallelogram, and the triangles defined by D are congruent:

$$\triangle PQR \cong \triangle RSP$$

Let E be a line segment connecting Q and S (Fig. 9-22B). Then E is a *major diagonal* of the parallelogram, and the triangles defined by E are congruent:

$$\triangle QRS \cong \triangle SPQ$$

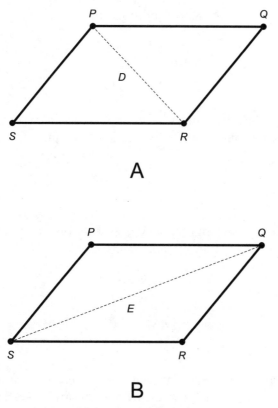

Fig. 9-22.   Triangles defined by the minor diagonal (A) or the major diagonal (B) of a parallelogram are congruent.

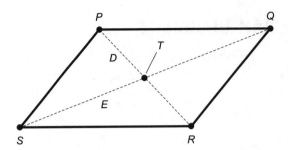

Fig. 9-23. The diagonals of a parallelogram bisect each other.

BISECTION OF PARALLELOGRAM DIAGONALS

Suppose we have a parallelogram defined by four distinct points P, Q, R, and S. Let D be the straight diagonal connecting P and R; let E be the straight diagonal connecting Q and S (Fig. 9-23). Then D and E bisect each other at their intersection point T. In addition, the following pairs of triangles are congruent:

$$\triangle PQT \cong \triangle RST$$
$$\triangle QRT \cong \triangle SPT$$

The converse of the foregoing is also true: if we have a plane quadrilateral whose diagonals bisect each other, then that quadrilateral is a parallelogram.

RECTANGLE

Suppose we have a parallelogram defined by four distinct points P, Q, R, and S. Suppose any of the following statements is true:

$$\angle PQR = 90°$$
$$\angle QRS = 90°$$
$$\angle RSP = 90°$$
$$\angle SPQ = 90°$$

Then all four interior angles measure $90°$, and the parallelogram is a *rectangle*: a four-sided plane polygon whose interior angles are all congruent (Fig. 9-24). The converse of this is also true: if a quadrilateral is a rectangle, then any given interior angle has a measure of $90°$.

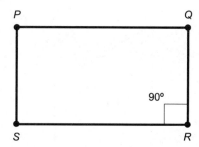

Fig. 9-24.   If a parallelogram has one right interior angle, then the parallelogram is a rectangle.

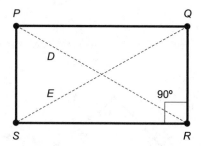

Fig. 9-25.   The diagonals of a rectangle have equal length.

RECTANGLE DIAGONALS

Imagine a parallelogram defined by four distinct points P, Q, R, and S. Let D be the straight diagonal connecting P and R; let E be the straight diagonal connecting Q and S. Let the length of D be denoted by d; let the length of E be denoted by e (Fig. 9-25). If $d=e$, then the parallelogram is a rectangle. The converse is also true: if a parallelogram is a rectangle, then $d=e$. A parallelogram is a rectangle if and only if its diagonals have equal lengths.

RHOMBUS DIAGONALS

Imagine a parallelogram defined by four distinct points P, Q, R, and S. Let D be the straight diagonal connecting P and R; let E be the straight diagonal connecting Q and S. If D is perpendicular to E, then the parallelogram is a *rhombus*, which is a four-sided plane polygon whose sides are all equally long (Fig. 9-26). The converse is also true: if a parallelogram is a rhombus, then D is perpendicular to E. A parallelogram is a rhombus if and only if its diagonals intersect at a right angle.

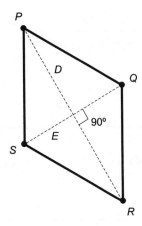

Fig. 9-26. The diagonals of a rhombus are perpendicular.

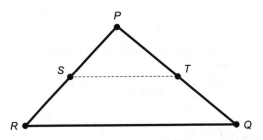

Fig. 9-27. A trapezoid within a triangle.

TRAPEZOID WITHIN TRIANGLE

Imagine a triangle defined by three distinct points P, Q, and R. Let S be the midpoint of side PR, and let T be the midpoint of side PQ. Then the straight line segments ST and RQ are parallel, and the figure defined by $STQR$ is a *trapezoid*: a four-sided plane polygon with one pair of parallel sides (Fig. 9-27). In addition, the length of line segment ST is half the length of line segment RQ.

MEDIAN OF A TRAPEZOID

Suppose we have a trapezoid defined by four distinct points P, Q, R, and S. Let T be the midpoint of side PS, and let U be the midpoint of side QR. Line segment TU is called the *median* of trapezoid $PQRS$. Let M be the polygon defined by P, Q, U, and T; let N be the polygon defined by T, U, R, and S. These two polygons, M and N, are trapezoids, as shown in Fig. 9-28.

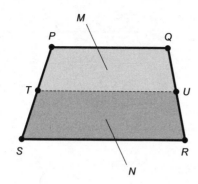

Fig. 9-28.   The median of a trapezoid.

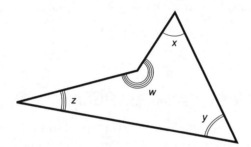

Fig. 9-29.   Interior angles of a plane quadrilateral.

In addition, the length of line segment TU is half the sum of the lengths of line segments PQ and SR. That is, the length of TU is equal to the average (arithmetic mean) of the lengths of PQ and SR.

SUM OF INTERIOR ANGLES OF PLANE QUADRILATERAL

Suppose we have a general plane quadrilateral, with no special properties other than the fact that it is defined by four distinct vertex points that all lie in the same plane. Let the measures of the interior angles be w, x, y, and z, as shown in Fig. 9-29. The following equation holds if the angular measures are given in degrees:

$$w + x + y + z = 360°$$

PERIMETER OF PARALLELOGRAM

Suppose we have a parallelogram defined by four distinct points P, Q, R, and S, in which two adjacent sides have lengths d and e as shown in Fig. 9-30.

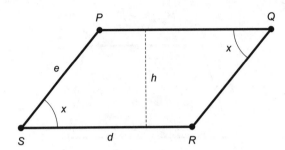

Fig. 9-30.   Perimeter and area of a parallelogram. If $d=e$, the figure is a rhombus.

Let d be the base length and let h be the height. Then the perimeter, B, of the parallelogram can be found using this formula:

$$B = 2d + 2e$$

INTERIOR AREA OF PARALLELOGRAM

Suppose we have a parallelogram as defined above and in Fig. 9-30. The interior area, A, of the parallelogram can be found using this formula:

$$A = dh$$

PERIMETER OF RHOMBUS

Imagine a rhombus defined by four distinct points P, Q, R, and S. All four sides of a rhombus have equal lengths. Therefore, the rhombus is a special case of the parallelogram (Fig. 9-30), in which $d=e$. Suppose that the lengths of all four sides are equal to d. The perimeter, B, of this rhombus is given by the following formula:

$$B = 4d$$

INTERIOR AREA OF RHOMBUS

Suppose we have a rhombus as defined above and in Fig. 9-30. The interior area, A, of the rhombus is given by:

$$A = dh$$

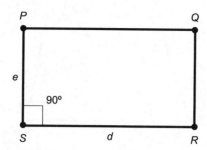

Fig. 9-31.   Perimeter and area of a rectangle. If $d=e$, the figure is a square.

PERIMETER OF RECTANGLE

Suppose we have a rectangle defined by four distinct points P, Q, R, and S, in which two adjacent sides have lengths d and e as shown in Fig. 9-31. Let d be the base length, and let e be the height. Then the perimeter, B, of the rectangle is given by the following formula:

$$B = 2d + 2e$$

INTERIOR AREA OF RECTANGLE

Suppose we have a rectangle as defined above and in Fig. 9-31. The interior area, A, is given by:

$$A = de$$

PERIMETER OF SQUARE

Suppose we have a square defined by four distinct points P, Q, R, and S, and having sides all of which have the same length. The square is a special case of the rectangle (Fig. 9-31), in which $d=e$. Suppose the lengths of all four sides are equal to d. The perimeter, B, of this square is given by the following formula:

$$B = 4d$$

INTERIOR AREA OF SQUARE

Suppose we have a square as defined above and in Fig. 9-31. The interior area, A, is given by:

$$A = d^2$$

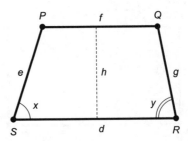

Fig. 9-32. Perimeter and area of a trapezoid.

PERIMETER OF TRAPEZOID

Imagine a trapezoid defined by four distinct points P, Q, R, and S, and having sides of lengths d, e, f, and g as shown in Fig. 9-32. Let d be the base length, let h be the height, let x be the measure of the angle between the sides having length d and e, and let y be the measure of the angle between the sides having length d and g. Suppose the sides having length d and f (line segments RS and PQ) are parallel. Then the perimeter, B, of the trapezoid is:

$$B = d + e + f + g$$

INTERIOR AREA OF TRAPEZOID

Suppose we have a trapezoid as defined above and in Fig. 9-32. The interior area, A, is equal to the product of the height and the average of the lengths of the base and the top. Here is the formula:

$$A = (dh + fh)/2$$

PROBLEM 9-5

Suppose a room has a north-facing wall that is rectangular, and that has a rectangular window in the center. The room measures 5.00 m from east-to-west, and the ceiling is 2.60 m high. The window, measured at its outer frame, measures 1.45 m high by 1.00 m wide. What is the surface area of the wall, not including the window?

SOLUTION 9-5

First, determine the area of the wall including the area of the window. If we call this area A_{wall}, then

$$A_{wall} = 5.00 \times 2.60$$
$$= 13.00 \, \text{m}^2$$

The area of the window (call it A_{win}), measured at its outer frame, is:

$$A_{win} = 1.45 \times 1.00$$

$$= 1.45 \text{ m}^2$$

Let A be the area of the wall, not including the window. Then

$$A = A_{wall} - A_{win}$$

$$= 13.00 - 1.45$$

$$= 11.55 \text{ m}^2$$

PROBLEM 9-6

Suppose a plot of land is shaped like a parallelogram. Adjacent sides measure 500 m and 400 m. What is the perimeter of this plot of land? What is its interior area?

SOLUTION 9-6

The perimeter of the plot of land is equal to the sum of twice the lengths of the adjacent sides. Suppose we call the lengths of the sides d and e, where $d = 500$ m and $e = 400$ m. Then the perimeter B, in meters, is:

$$B = 2d + 2e$$

$$= 2 \times 500 + 2 \times 400$$

$$= 1000 + 800$$

$$= 1800 \text{ m}$$

In order to calculate the area of the field, we must know how wide it is. For fixed lengths of sides, the width of a parallelogram depends on the angle between two adjacent sides (in this case, boundaries of the field). We aren't given this information, so we can't determine the area of the field.

Circles and Ellipses

That's enough of figures with straight lines. Let's get into plane curves. In some ways, the following formulas are easier for mathematicians to derive than the ones for figures consisting of lines and angles.

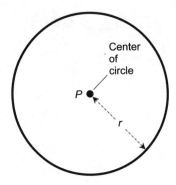

Fig. 9-33.   Perimeter and area of a circle.

PERIMETER OF CIRCLE

Suppose we have a circle having radius r as shown in Fig. 9-33. Then the perimeter, B, also called the *circumference*, of the circle is given by the following formula:

$$B = 2\pi r$$

INTERIOR AREA OF CIRCLE

Suppose we have a circle as defined above and in Fig. 9-33. The interior area, A, of the circle can be found using this formula:

$$A = \pi r^2$$

INTERIOR AREA OF ELLIPSE

Suppose we have an ellipse whose *major half-axis* (the longer half-axis or largest radius) measures r, and whose *minor half-axis* (the shorter half-axis or smallest radius) measures s as shown in Fig. 9-34. The interior area, A, of the ellipse is given by:

$$A = \pi r s$$

PROBLEM 9-7
What is the circumference of a circle whose radius is 10.0000 m? What is its interior area?

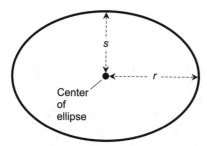

Fig. 9-34. Interior area of an ellipse.

SOLUTION 9-7

In this case, $r = 10.0000$, expressed in meters. To calculate the circumference B in meters, proceed as follows. Consider the value of π to be 3.14159.

$$B = 2\pi r$$
$$= 2 \times 3.14159 \times 10.0000$$
$$= 62.8318 \text{ m}$$

To calculate the interior area A in square meters, proceed as follows:

$$A = \pi r^2$$
$$= 3.14159 \times 10^2$$
$$= 314.59 \text{ m}^2$$

PROBLEM 9-8

Consider an ellipse whose minor half-axis is 10.0000 m and whose major half-axis can be varied at will. What must the major half-axis be in order to get an interior area of 200.000 m^2? Consider the value of π to be 3.14159.

SOLUTION 9-8

In this problem, $A = 200$ and $r = 10$. We seek to find the value of s in this formula:

$$A = \pi r s$$

Substituting the values in, the calculation goes like this:

$$200.000 = 3.14159 \times 10.0000 \times s$$
$$10.0000 \times s = 200.000/3.14159$$
$$s = 6.36620 \text{ m}$$

Quiz

Refer to the text in this chapter if necessary. A good score is eight correct. Answers are in the back of the book.

1. In an equilateral plane triangle, each interior angle
    (a) has a measure less than 60°
    (b) has a measure greater than 60°
    (c) has a measure of 60°
    (d) has a measure different from either of the other two angles

2. Suppose a building is constructed in the shape of a parallelogram, rather than in the usual rectangular shape, because of the strange layout of streets in that part of town. The sides of the building run east–west and more or less northeast–southwest. The length of the building measured along the east–west walls is 30 m, and the distance between the northerly east–west end and the southerly east–west end is 40 m. What is the area of the basement floor (including the area taken up by interior and exterior walls)?
    (a) $1500 \, m^2$
    (b) $1200 \, m^2$
    (c) $750 \, m^2$
    (d) More information is needed to answer this question.

3. Suppose a building is constructed in the shape of a parallelogram, rather than in the usual rectangular shape, because of the strange layout of streets in that part of town. The sides of the building run east–west and more or less northeast–southwest. The length of the building measured along the east–west walls is 30 m, and the length measured along the northeast–southwest walls is 40 m. What is the area of the basement floor (including the area taken up by interior and exterior walls)?
    (a) $1500 \, m^2$
    (b) $1200 \, m^2$
    (c) $750 \, m^2$
    (d) More information is needed to answer this question.

4. If the radius of a circle is quadrupled, what happens to its interior area?
    (a) It becomes twice as great.
    (b) It becomes four times as great.
    (c) It becomes 16 times as great.
    (d) More information is needed to answer this question.

5. If the length of the minor semi-axis of an ellipse is quadrupled while the length of the major semi-axis is cut in half, what happens to the interior area?
    (a) It stays the same.
    (b) It increases by a factor of the square root of 2.
    (c) It increases by a factor of 2.
    (d) It increases by a factor of 4.

6. A right plane triangle cannot
    (a) be an isosceles triangle
    (b) have three sides all of different lengths
    (c) have three sides all of the same length
    (d) have any interior angle measuring less than 90°

7. Suppose three towns, called Jimsville, Joesville, and Johnsville, all lie along a perfectly straight, east–west stretch of highway called Route 999. As you drive from west to east, you encounter Jimsville first, then Joesville, and then Johnsville. Suppose the distance between Jimsville and Joesville is 8 kilometers (km), and the distance between Joesville and Johnsville is 6 km. What is the distance between Jimsville and Johnsville as measured along Route 999?
    (a) 8 km
    (b) 10 km
    (c) 14 km
    (d) More information is needed to answer this question.

8. Suppose Jimsville and Johnsville lie along a perfectly straight, east–west stretch of highway called Route 999. Joesville lies north of Route 999, and is in a position such that it is 8 km away from Jimsville and 6 km away from Johnsville. Suppose a straight line connecting Jimsville and Joesville is perpendicular to a straight line connecting Johnsville and Joesville. What is the distance between Jimsville and Johnsville as measured along Route 999?
    (a) 8 km
    (b) 10 km
    (c) 14 km
    (d) More information is needed to answer this question.

9. Suppose Jimsville and Johnsville lie along a perfectly straight, east–west stretch of highway called Route 999. Joesville lies south of Route 999, and is in a position such that it is 8 km away from Jimsville and 8 km away from Johnsville. All three towns lie at the

vertices of an isosceles triangle. What is the distance between Jimsville and Johnsville as measured along Route 999?

(a) 8 km

(b) 10 km

(c) 14 km

(d) More information is needed to answer this question.

10. Suppose Jimsville and Johnsville lie along a perfectly straight, east–west stretch of highway called Route 999. Joesville lies south of Route 999, and is in a position such that it is 8 km away from Jimsville and 8 km away from Johnsville. All three towns lie at the vertices of an equilateral triangle. What is the distance between Jimsville and Johnsville as measured along Route 999?

(a) 8 km

(b) 10 km

(c) 14 km

(d) More information is needed to answer this question.

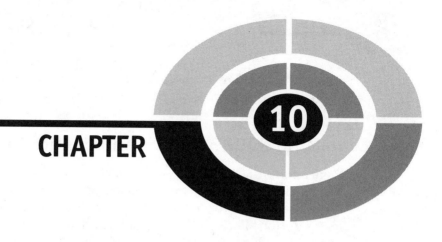

Geometry in Space

Solid geometry involves points, lines, and planes. The difference between two-dimensional (2D) geometry and three-dimensional (3D) geometry is the fact that, well, there's an extra dimension! This makes things more interesting, and also more complicated.

Points, Lines, and Planes

A point can be envisioned as an infinitely tiny sphere, having height, width, and depth all equal to zero, but nevertheless possessing a specific location. A point is *zero-dimensional* (0D). A line can be thought of as an infinitely thin, perfectly straight, infinitely long thread or wire. It is *one-dimensional* (1D). A *plane* can be imagined as an infinitely thin, perfectly flat surface having an infinite expanse. It is *two-dimensional* (2D). *Space* in the simplest sense is the set of points representing all possible physical locations in the universe at any specific instant in time. Space is *three-dimensional* (3D).

If time is included in a concept of space, we get *four-dimensional* (4D) space, also known as *hyperspace*.

NAMING POINTS, LINES, AND PLANES

Points, lines, and planes in solid geometry are usually named using upper-case, italicized letters of the alphabet, just as they are in plane geometry. A common name for a point is *P* (for "point"). A common name for a line is *L* (for "line"). When it comes to planes in 3D space, we must use our imaginations. The letters *X*, *Y*, and *Z* are good choices. Sometimes lower-case, non-italic letters are used, such as m and n.

When we have two or more points, the letters immediately following *P* are used, for example *Q*, *R*, and *S*. If two or more lines exist, the letters immediately following *L* are used, for example *M* and *N*. Alternatively, numeric subscripts can be used. We might have points called P_1, P_2, P_3, and so forth, lines called L_1, L_2, L_3, and so forth, and planes called X_1, X_2, X_3, and so forth.

THREE POINT PRINCIPLE

Suppose that *P*, *Q*, and *R* are three different geometric points, no two of which lie on the same line. Then these points define one and only one (a unique or specific) plane *X*. The following two statements are always true, as shown in Fig. 10-1:

- *P*, *Q*, and *R* lie in a common plane *X*.
- *X* is the only plane in which all three points lie.

In order to show that a surface extends infinitely in 2D, we have to be imaginative. It's not as easy as showing that a line extends infinitely in 1D, because there aren't any good ways to draw arrows on the edges of a plane region the way we can draw them on the ends of a line segment.

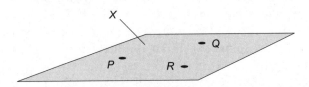

Fig 10-1. Three points *P*, *Q*, and *R*, not all on the same line, define a specific plane *X*. The plane extends infinitely in 2D.

It is customary to draw planes as rectangles in perspective; they appear as rectangles, parallelograms, or trapezoids when rendered on a flat page. This is all right, as long as it is understood that the surface extends infinitely in 2D.

INTERSECTING LINE PRINCIPLE

Suppose that lines L and M intersect in a point P. Then the two lines define a unique plane X. The following two statements are always true, as shown in Fig. 10-2:

- L and M lie in a common plane X.
- X is the only plane in which both lines lie.

LINE AND POINT PRINCIPLE

Let L be a line and P be a point not on that line. Then line L and point P define a unique plane X. The following two statements are always true:

- L and P lie in a common plane X.
- X is the only plane in which both L and P lie.

PLANE REGIONS

The 2D counterpart of the 1D line segment is the *simple plane region*. A simple plane region consists of all the points inside a polygon or enclosed curve. The polygon or curve itself might be included in the set of points comprising the simple plane region, but this is not necessarily the case. If the polygon or curve is included, the simple plane region is said to be *closed*. Some examples are denoted in Fig. 10-3A; the boundaries are drawn so they look continuous. If the polygon or curve is not included, the simple plane

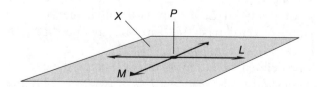

Fig. 10-2.   Two lines L and M, intersecting at point P, define a specific plane X. The plane extends infinitely in 2D.

Fig. 10-3.   Plane regions. At A, closed; at B, open.

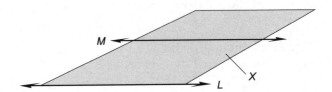

Fig. 10-4.   A half plane X, defined by two parallel lines, L and M. The half plane extends infinitely in 2D on the "M" side of L.

region is said to be *open*. In Fig. 10-3B, several examples of open simple plane regions are denoted; the boundaries are dashed.

The respective regions in Figs. 10-3A and B have identical shapes. They also have identical perimeters and identical interior areas. The outer boundaries do not add anything to the perimeter or the interior area.

These examples show specialized cases, in which the regions are contiguous, or "all of a piece," and the boundaries are either closed all the way around or open all the way around. Some plane regions have boundaries that are closed part of the way around, or in segments; it is also possible to have plane regions composed of two or more non-contiguous subregions.

HALF PLANES

Sometimes, it is useful to talk about the portion of a geometric plane that lies "on one side" of a certain line. In Fig. 10-4, imagine the union of all possible geometric rays that start at L, then pass through line M (which is parallel to L), and extend onward past M "forever" in one direction. The region thus defined is known as a *half plane*.

The half plane defined by L and M might include the end line L, in which case it is *closed-ended*. In this case, line L is drawn as a solid line, as shown in Fig. 10-4. But the end line might not comprise part of the half plane, in which case the half plane is *open-ended*. Then line L is drawn as a dashed line.

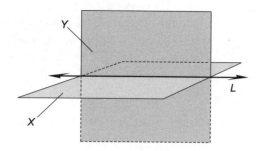

Fig. 10-5. The intersection of two planes X and Y determines a unique line L. The planes extend infinitely in 2D.

INTERSECTING PLANES

Suppose that two different planes X and Y intersect; that is, they have points in common. Then the two planes intersect in a unique line L. The following two statements are always true, as shown in Fig. 10-5:

- Planes X and Y share a common line L.
- L is the only line that planes X and Y have in common.

PARALLEL LINES IN 3D SPACE

By definition, two different lines L and M in three-space are *parallel lines* if and only if both of the following are true:

- Lines L and M do not intersect.
- Lines L and M lie in the same plane X.

If two lines are parallel and they lie in a given plane X, then X is the only plane in which the two lines lie. Thus, two parallel lines define a unique plane in Euclidean space.

SKEW LINES

By definition, two lines L and M in three-space are *skew lines* if and only if both of the following are true:

- Lines L and M do not intersect.
- Lines L and M do not lie in the same plane.

Imagine an infinitely long, straight, two-lane highway and an infinitely long, straight power line propped up on utility poles. Further imagine that the power line and the highway center line are both infinitely thin, and that the power line doesn't sag between the poles. Suppose the power line passes over the highway somewhere. Then the center line of the highway and the power line define skew lines.

PROBLEM 10-1
Find an example of a theoretical plane region with a finite, nonzero area but an infinite perimeter.

SOLUTION 10-1
Examine Fig. 10-6. Suppose the three lines PQ, RS, and TU (none of which are part of the plane region X, but are shown only for reference) are mutually parallel, and that the distances d_1, d_2, d_3, . . . are such that $d_2 = d_1/2$, $d_3 = d_2/2$, and in general, for any positive integer n, $d_{(n+1)} = d_n/2$. Also suppose that the length of line segment PV is greater than the length of line segment PT. Then plane region X has an infinite number of sides, each of which has a length greater than the length of line segment PT, so its perimeter is infinite. But the interior area of X is finite and nonzero, because it is obviously less than the interior area of quadrilateral $PQSR$ but greater than the area of quadrilateral $TUSR$.

PROBLEM 10-2
How many planes can mutually intersect in a given line L?

SOLUTION 10-2
In theory, an infinite number of planes can all intersect in a common line. Think of the line as an infinitely long hinge, and then imagine a plane that can swing around this hinge. Each position of the swinging plane represents a unique plane in space.

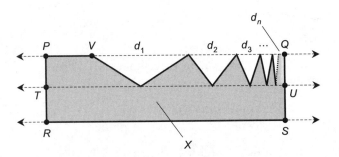

Fig. 10-6. Illustration for Problem 10-1.

Straight-Edged Objects

In Euclidean three-space, geometric solids with straight edges have flat faces, also called *facets*, each of which forms a plane polygon. An object of this sort is known as a *polyhedron*.

THE TETRAHEDRON

A polyhedron in 3D must have at least four faces. A four-faced polyhedron is called a *tetrahedron*. Each of the four faces of a tetrahedron is a triangle. There are four vertices. Any four specific points, not all in a single plane, form a unique tetrahedron.

SURFACE AREA OF TETRAHEDRON

Figure 10-7 shows a tetrahedron. The surface area is found by adding up the interior areas of all four triangular faces. In the case of a *regular tetrahedron*, all six edges have the same length, and therefore all four faces are equilateral triangles. If the length of each edge of a regular tetrahedron is equal to s units, then the surface area, B, of the whole four-faced regular tetrahedron is given by:

$$B = 3^{1/2}s^2$$

In this formula, $3^{1/2}$ represents the square root of 3, or approximately 1.732.

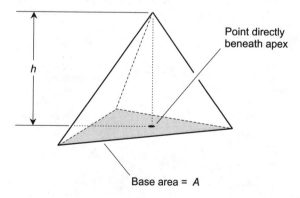

Fig. 10-7.   A tetrahedron has four faces (including the base) and six edges.

VOLUME OF TETRAHEDRON

Imagine a tetrahedron whose base is a triangle with area A, and whose height is h as shown in Fig. 10-7. The volume, V, of the figure is given by:

$$V = Ah/3$$

PYRAMID

Figure 10-8 illustrates a *pyramid*. This figure has a square or rectangular base and four slanted faces. If the base is a square, and if the *apex* (the top of the pyramid) lies directly above a point at the center of the base, then the figure is a *regular pyramid*, and all of the slanted faces are isosceles triangles.

SURFACE AREA OF PYRAMID

The surface area of a pyramid is found by adding up the areas of all five of its faces (the four slanted faces plus the base). In the case of a regular pyramid where the length of each slanted edge, called the *slant height*, is equal to s units and the length of each edge of the base is equal to t units, the surface area, B, is given by:

$$B = t^2 + 2t(s^2 - t^2/4)^{1/2}$$

In the case of an *irregular pyramid*, the problem of finding the surface area is more complicated, because it involves individually calculating the area of the base and each slanted face, and then adding all the areas up.

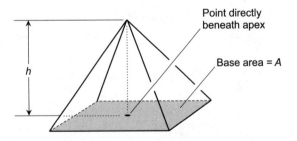

Point directly beneath apex

Base area = A

h

Fig. 10-8. A pyramid has five faces (including the base) and eight edges.

VOLUME OF PYRAMID

Imagine a pyramid whose base is a square with area A, and whose height is h as shown in Fig. 10-8. The volume, V, of the pyramid is given by:

$$V = Ah/3$$

This is true whether the pyramid is regular or irregular.

THE CUBE

Figure 10-9 illustrates a *cube*. This is a *regular hexahedron* (six-sided polyhedron). It has 12 edges, each of which is of the same length, and eight vertices. Each of the six faces is a square.

SURFACE AREA OF CUBE

Imagine a cube whose edges each have length s, as shown in Fig. 10-9. The surface area, A, of the cube is given by:

$$A = 6s^2$$

VOLUME OF CUBE

Imagine a cube as defined above and in Fig. 10-9. The volume, V, of the solid enclosed by the cube is given by:

$$V = s^3$$

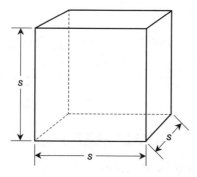

Fig. 10-9.   A cube has six square faces and 12 edges of identical length.

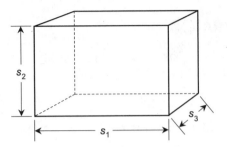

Fig. 10-10.   A rectangular prism has six rectangular faces and 12 edges.

THE RECTANGULAR PRISM

Figure 10-10 illustrates a *rectangular prism*. This is a hexahedron, each of whose six faces is a rectangle. The figure has 12 edges and eight vertices.

SURFACE AREA OF RECTANGULAR PRISM

Imagine a rectangular prism whose edges have lengths s_1, s_2, and s_3 as shown in Fig. 10-10. The surface area, A, of the prism is given by:

$$A = 2s_1s_2 + 2s_1s_3 + 2s_2s_3$$

VOLUME OF RECTANGULAR PRISM

Imagine a rectangular prism as defined above and in Fig. 10-10. The volume, V, of the enclosed solid is given by:

$$V = s_1s_2s_3$$

THE PARALLELEPIPED

A *parallelepiped* is a six-faced polyhedron in which each face is a parallelogram, and opposite pairs of faces are congruent. The figure has 12 edges and eight vertices. The acute angles between the pairs of edges are x, y, and z, as shown in Fig. 10-11.

SURFACE AREA OF PARALLELEPIPED

Imagine a parallelepiped with faces of lengths s_1, s_2, and s_3. Suppose the angles between pairs of edges are x, y, and z as shown in Fig. 10-11.

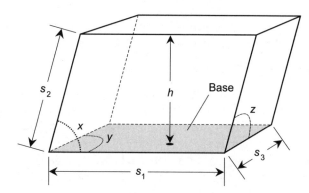

Fig. 10-11.   A parallelepiped has six faces, all of which are parallelograms, and 12 edges.

The surface area, A, of the parallelepiped is given by:

$$A = 2s_1s_2 \sin x + 2s_1s_3 \sin y + 2s_2s_3 \sin z$$

where $\sin x$ represents the sine of angle x, $\sin y$ represents the sine of angle y, and $\sin z$ represents the sine of angle z. Sines of angles can be found with any scientific calculator, including the ones in most personal computers.

VOLUME OF PARALLELEPIPED

Imagine a parallelepiped whose faces have lengths s_1, s_2, and s_3, and that has angles between edges of x, y, and z as shown in Fig. 10-11. Suppose further that the height of the parallelepiped, as measured along a line normal to the base, is equal to h. The volume, V, of the enclosed solid is equal to the product of the base area and the height:

$$V = hs_1s_3 \sin y$$

PROBLEM 10-3
Suppose you want to paint the interior walls of a room in a house. The room is shaped like a rectangular prism. The ceiling is exactly 3.0 m above the floor. The floor and the ceiling both measure exactly 4.2 m by 5.5 m. There are two windows, the outer frames of which both measure 1.5 m high by 1.0 m wide. There is one doorway, the outer frame of which measures 2.5 m high by 1.0 m wide. With two coats of paint (which you intend to apply), one liter of paint can be expected to cover exactly 20 m² of wall area. How much paint, in liters, will you need to completely do the job? Note that a liter is defined

as a cubic decimeter, or the volume of a cube measuring 0.1 meter on an edge. This means that a liter is equal to 0.001 cubic meter (m^3).

SOLUTION 10-3

It is necessary to find the amount of wall area that this room has. Based on the information given, we can conclude that the rectangular prism formed by the edges between walls, floor, and ceiling measures 3.0 m high by 4.2 m wide by 5.5 m deep. So we can let $s_1 = 3.0$, $s_2 = 4.2$, and $s_3 = 5.5$ (with all units assumed to be in meters) to find the surface area A of the rectangular prism, in square meters, neglecting the area subtracted for the windows and doorway. Using the formula:

$$A = 2s_1s_2 + 2s_1s_3 + 2s_2s_3$$
$$= (2 \times 3.0 \times 4.2) + (2 \times 3.0 \times 5.5) + (2 \times 4.2 \times 5.5)$$
$$= 25.2 + 33.0 + 46.2$$
$$= 104.4 \, m^2$$

There are two windows measuring 1.5 m by 1.0 m. Each window takes away $1.5 \times 1.0 = 1.5 \, m^2$ of area. The doorway measures 2.5 m by 1.0 m, so it takes away $2.5 \times 1.0 = 2.5 \, m^2$. Therefore, the windows and doorway combined take away $1.5 + 1.5 + 2.5 = 5.5 \, m^2$ of wall space. We must also take away the areas of the floor and ceiling. This is the final factor in the above equation, $2s_2s_3 = 46.2$. The wall area to be painted, call it A_w, is calculated this way:

$$A_w = (104.4 - 5.5) - 46.2$$
$$= 52.7 \, m^2$$

A liter of paint can be expected to cover $20 \, m^2$. So we will need 52.7/20, or 2.635, liters of paint to do this job.

Cones, Cylinders, and Spheres

A *cone* has a circular or elliptical base and an apex point. The cone itself consists of the union of the following sets of points:

- The circle or ellipse.
- All points inside the circle or ellipse and that lie in its plane.
- All line segments connecting the circle or ellipse (not including its interior) and the apex point.

The interior of the cone consists of the set of all points within the cone. The surface might or might not be included in the definition of the interior.

A *cylinder* has a circular or elliptical base, and a circular or elliptical top that is congruent to the base and that lies in a plane parallel to the base. The cylinder itself consists of the union of the following sets of points:

- The base circle or ellipse.
- All points inside the base circle or ellipse and that lie in its plane.
- The top circle or ellipse.
- All points inside the top circle or ellipse and that lie in its plane.
- All line segments connecting corresponding points on the base circle or ellipse and top circle or ellipse (not including their interiors).

The interior of the cylinder consists of the set of all points within the cylinder. The surface might or might not be included in the definition of the interior.

These are general definitions, and they encompass a great variety of objects! Here, we'll look only at cones and cylinders whose bases are circles.

THE RIGHT CIRCULAR CONE

A *right circular cone* has a base that is a circle, and an apex point on a line perpendicular to the base and passing through its center. This type of cone, when the height is about twice the diameter of the base, has a "dunce cap" shape. Dry sand, when poured into an enormous pile on a flat surface, acquires a shape that is roughly that of a right circular cone. An example of this type of object is shown in Fig. 10-12.

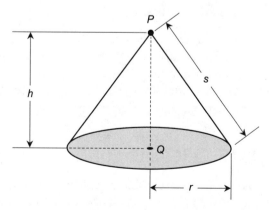

Fig. 10-12.   A right circular cone.

SURFACE AREA OF RIGHT CIRCULAR CONE

Imagine a right circular cone as shown in Fig. 10-12. Let P be the apex of the cone, and let Q be the center of the base. Let r be the radius of the base, let h be the height of the cone (the length of line segment PQ), and let s be the slant height of the cone as measured from any point on the edge of the base to the apex P. The surface area S_1 of the cone, including the base, is given by either of the following formulas:

$$S_1 = \pi r^2 + \pi r s$$
$$S_1 = \pi r^2 + \pi r (r^2 + h^2)^{1/2}$$

The surface area S_2 of the cone, not including the base, is called the *lateral surface area* and is given by either of the following:

$$S_2 = \pi r s$$
$$S_2 = \pi r (r^2 + h^2)^{1/2}$$

VOLUME OF RIGHT CIRCULAR CONE

Imagine a right circular cone as defined above and in Fig. 10-12. The volume, V, of the interior of the figure is given by:

$$V = \pi r^2 h / 3$$

THE SLANT CIRCULAR CONE

A *slant circular cone* has a base that is a circle, just as does the right circular cone. But the apex is such that a line, perpendicular to the plane containing the base and running from the apex through the plane containing the base, does not pass through the center of the base. Such a cone has a "blown-over" or "cantilevered" appearance (Fig. 10-13).

VOLUME OF SLANT CIRCULAR CONE

Imagine a cone whose base is a circle. Let P be the apex of the cone, and let Q be a point in the plane X containing the base such that line segment PQ is perpendicular to X (Fig. 10-13). Let h be the height of the cone (the length

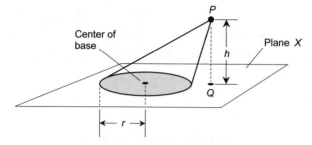

Fig. 10-13.   A slant circular cone.

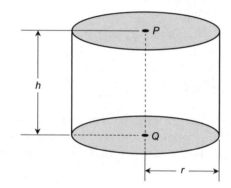

Fig. 10-14.   A right circular cylinder.

of line segment PQ). Let r be the radius of the base. Then the volume, V, of the corresponding cone is given by:

$$V = \pi r^2 h/3$$

This is the same as the formula for the volume of a right circular cone.

THE RIGHT CIRCULAR CYLINDER

A *right circular cylinder* has a circular base and a circular top. The base and the top lie in parallel planes. The center of the base and the center of the top lie along a line that is perpendicular to both the plane containing the base and the plane containing the top. A general example is shown in Fig. 10-14.

SURFACE AREA OF RIGHT CIRCULAR CYLINDER

Imagine a right circular cylinder where P is the center of the top and Q is the center of the base (Fig. 10-14). Let r be the radius of the cylinder, and let h be the height (the length of line segment PQ). Then the surface area S_1 of the cylinder, including the base, is given by:

$$S_1 = 2\pi rh + 2\pi r^2 = 2\pi r(h + r)$$

The lateral surface area S_2 of the cylinder (not including the base) is given by:

$$S_2 = 2\pi rh$$

VOLUME OF RIGHT CIRCULAR CYLINDER

Imagine a right circular cylinder as defined above and shown in Fig. 10-14. The volume, V, of the cylinder is given by:

$$V = \pi r^2 h$$

THE SLANT CIRCULAR CYLINDER

A *slant circular cylinder* has a circular base and a circular top. The base and the top lie in parallel planes. The center of the base and the center of the top lie along a line that is not perpendicular to the planes that contain them (Fig. 10-15).

VOLUME OF SLANT CIRCULAR CYLINDER

Imagine a slant circular cylinder as defined above and in Fig. 10-15. The volume, V, of the corresponding solid is given by:

$$V = \pi r^2 h$$

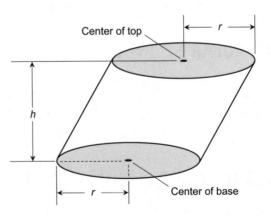

Fig. 10-15.　A slant circular cylinder.

THE SPHERE

Consider a specific point P in 3D space. The surface of a sphere S consists of the set of all points at a specific distance or radius r from point P. The interior of sphere S, including the surface, consists of the set of all points whose distance from point P is less than or equal to r. The interior of sphere S, not including the surface, consists of the set of all points whose distance from P is less than r.

SURFACE AREA OF SPHERE

Imagine a sphere S having radius r as shown in Fig. 10-16. The surface area, A, of the sphere is given by:

$$A = 4\pi r^2$$

VOLUME OF SPHERE

Imagine a sphere S as defined above and in Fig. 10-16. The volume, V, of the solid enclosed by the sphere is given by:

$$V = 4\pi r^3/3$$

This volume applies to the interior of sphere S, either including the surface or not including it, because the surface has zero volume.

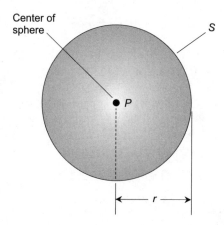

Fig. 10-16.   A sphere.

PROBLEM 10-4

A cylindrical water tower is exactly 30 m high and exactly 10 m in radius. How many liters of water can it hold, assuming the entire interior can be filled with water? Round the answer off to the nearest liter.

SOLUTION 10-4

Use the formula for the volume of a right circular cylinder to find the volume in cubic meters:

$$V = \pi r^2 h$$

Plug in the numbers. Let $r = 10$, $h = 30$, and $\pi = 3.14159$:

$$V = 3.14159 \times 10^2 \times 30$$
$$= 3.14159 \times 100 \times 30$$
$$= 9424.77 \, \text{m}^3$$

There are 1000 liters per cubic meter. This means that the amount of water the tower can hold, in liters, is equal to 9424.77 × 1000, or 9,424,770.

PROBLEM 10-5

A circus tent is shaped like a right circular cone. Its diameter is 50 m and the height at the center is 20 m. How much canvas is in the tent? Express the answer to the nearest square meter.

SOLUTION 10-5

Use the formula for the lateral surface area, S, of the right circular cone:

$$S = \pi r (r^2 + h^2)^{1/2}$$

We know that the diameter is 50 m, so the radius is 25 m. Therefore, $r = 25$. We also know that $h = 20$. Let $\pi = 3.14159$. Then:

$$S = 3.14159 \times 25 \times (25^2 + 20^2)^{1/2}$$
$$= 3.14159 \times 25 \times (625 + 400)^{1/2}$$
$$= 3.14159 \times 25 \times 1025^{1/2}$$
$$= 3.14159 \times 25 \times 32.0156$$
$$= 2514.4972201 \, \text{m}^2$$

There are 2514 m^2 of canvas, rounded off to the nearest square meter.

PROBLEM 10-6
Suppose a football field is to be covered by an inflatable dome that takes the shape of a half-sphere. If the radius of the dome is 100 m, what is the volume of air enclosed by the dome, in cubic meters? Find the result to the nearest 1000 m³.

SOLUTION 10-6
First, we must find the volume V of a sphere whose radius is 100 m, and then divide the result by 2. Let $\pi = 3.14159$. Using the formula with $r = 100$ gives this result:

$$V = (4 \times 3.14159 \times 100^3)/3$$
$$= (4 \times 3.14159 \times 1,000,000)/3$$
$$= 4,188,786.667 \, m^3$$

Thus $V/2 = 4,188,786.667/2 = 2,094,393.333 \, m^3$. Rounding off to the nearest 1000 m³, we obtain the figure of 2,094,000 m³ as the volume of air enclosed by the dome.

Quiz

Refer to the text in this chapter if necessary. A good score is eight correct. Answers are in the back of the book.

1. What is the approximate surface area of the sides (not including the base or top) of a water tower that is shaped like a right circular cylinder, that has a radius of exactly 5 meters, and that is exactly 20 meters tall?
   (a) 1571 m²
   (b) 628.3 m²
   (c) 100 m²
   (d) It cannot be determined without more information.

2. If the height of a silo that is shaped like a right circular cylinder is cut in half while its radius remains the same, what happens to its surface area (not including the base or the top)?
   (a) It decreases to 1/2 of its previous value.
   (b) It decreases to 1/4 of its previous value.
   (c) It decreases to 1/8 of its previous value.
   (d) It decreases to 1/16 of its previous value.

3. If the height of a silo that is shaped like a right circular cylinder is cut in half while its radius remains the same, what happens to its volume?
    (a) It decreases to 1/2 of its previous value.
    (b) It decreases to 1/4 of its previous value.
    (c) It decreases to 1/8 of its previous value.
    (d) It decreases to 1/16 of its previous value.

4. What is the approximate volume, in cubic centimeters (cm^3), of a ball that is exactly 20 cm in diameter? (Remember that the diameter of a sphere is twice the radius.)
    (a) 62.83 cm^3
    (b) 1257 cm^3
    (c) 4189 cm^3
    (d) It cannot be determined without more information.

5. Which of the following statements is not true in Euclidean 3D space?
    (a) If two straight lines are parallel, then they are skew.
    (b) If two straight lines are parallel, then they lie in the same plane.
    (c) If two straight lines are parallel, then they determine a unique plane.
    (d) If two straight lines are parallel, then they do not intersect.

6. Which of the following statements is always true in Euclidean 3D space?
    (a) Four points determine a unique flat plane.
    (b) Three points can have more than one flat plane in common.
    (c) Two lines that do not intersect determine a unique flat plane.
    (d) None of the statements (a), (b), or (c) is always true in Euclidean 3D space.

7. If the radius of a sphere increases by a factor of 10, the volume of the sphere increases by a factor of
    (a) 100
    (b) $(400/3)\pi$
    (c) 1000
    (d) $(4000/3)\pi$

8. The boundary of a closed plane region
    (a) adds nothing to its area
    (b) increases its volume
    (c) reduces its perimeter
    (d) is always a straight line

9. A cube is a special form of
   (a) rhombus
   (b) pyramid
   (c) cone
   (d) parallelepiped

10. If the radius of a sphere increases by a factor of 9, the ratio of the volume of the sphere to its surface area increases by a factor of
   (a) 3
   (b) 9
   (c) 27
   (d) 81

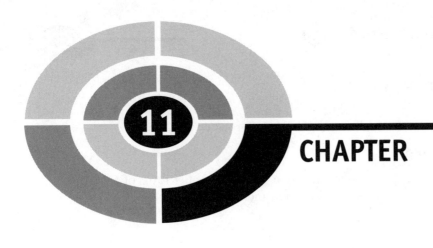

CHAPTER 11

Graphing It

We've already done some work with simple two-dimensional (2D) graphs. In this chapter, we'll take a closer look at the two most common coordinate systems on which 2D graphs are plotted. They are the *Cartesian plane* (named after the French mathematician René Descartes who invented it in the 1600s) and the *polar coordinate plane*.

The Cartesian Plane

The Cartesian plane, also called the *rectangular coordinate plane* or *rectangular coordinates*, is defined by two number lines that intersect at a right angle. Figure 11-1 illustrates a basic set of rectangular coordinates. Both number lines have equal increments. This means that on either axis, points corresponding to consecutive integers are the same distance apart, no matter where on the axis we look. The two number lines intersect at their zero points. The horizontal (right-and-left) axis is the *x axis*; the vertical (up-and-down) axis is the *y axis*.

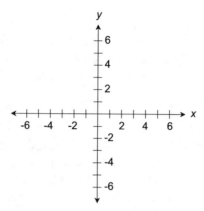

Fig. 11-1.   The Cartesian plane is defined by two number lines that intersect at right angles.

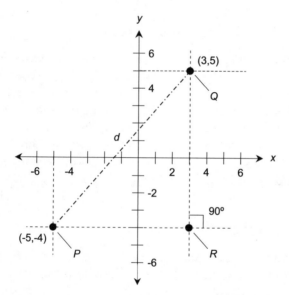

Fig. 11-2.   Two points P and Q, plotted in rectangular coordinates, and a third point R, important in finding the distance d between P and Q.

ORDERED PAIRS AS POINTS

Figure 11-2 shows two specific points P and Q on the Cartesian plane. Any given point on the plane can be denoted as an *ordered pair* in the form (x,y), determined by the numerical values at which perpendiculars from the point intersect the x and y axes. The coordinates of point P are $(-5,-4)$, and the coordinates of point Q are $(3,5)$. In Fig. 11-2, the perpendiculars are shown as horizontal and vertical dashed lines. When denoting an ordered pair, it is customary to place the two numbers or variables together right up against the comma; there is no space after the comma.

The word "ordered" means that the order in which the numbers are listed is important. The ordered pair (7,2) is not the same as the ordered pair (2,7), even though both pairs contain the same two numbers. In this respect, ordered pairs are different than mere sets of numbers. Think of a two-lane highway with a northbound lane and a southbound lane. If there is never any traffic on the highway, it doesn't matter which lane (the one on the eastern side or the one on the western side) is called "northbound" and which is called "southbound." But when there are vehicles on that road, it makes a big difference! The untraveled road is like a set; the traveled road is like an ordered pair.

ABSCISSA, ORDINATE, AND ORIGIN

In any graphing scheme, there is at least one *independent variable* and at least one *dependent variable*. You learned about these types of variables in Chapter 2. The independent-variable coordinate (usually x) of a point on the Cartesian plane is called the *abscissa*, and the dependent-variable coordinate (usually y) is called the *ordinate*. The point where the two axes intersect, in this case (0,0), is called the *origin*. In the scenario shown by Fig. 11-2, point P has an abscissa of -5 and an ordinate of -4, and point Q has an abscissa of 3 and an ordinate of 5.

DISTANCE BETWEEN POINTS

Suppose there are two different points $P=(x_0,y_0)$ and $Q=(x_1,y_1)$ on the Cartesian plane. The distance d between these two points can be found by determining the length of the hypotenuse, or longest side, of a right triangle PQR, where point R is the intersection of a "horizontal" line through P and a "vertical" line through Q. In this case, "horizontal" means "parallel to the x axis," and "vertical" means "parallel to the y axis." An example is shown in Fig. 11-2. Alternatively, we can use a "horizontal" line through Q and a "vertical" line through P to get the point R. The resulting right triangle in this case has the same hypotenuse, line segment PQ, as the triangle determined the other way.

The Pythagorean theorem from plane geometry states that the square of the length of the hypotenuse of a right triangle is equal to the sum of the squares of the other two sides. In the Cartesian plane, that means the following equation always holds true:

$$d^2 = (x_1 - x_0)^2 + (y_1 - y_0)^2$$

and therefore:

$$d = [(x_1 - x_0)^2 + (y_1 - y_0)^2]^{1/2}$$

where the 1/2 power is the square root. In the situation shown in Fig. 11-2, the distance d between points $P = (x_0, y_0) = (-5, -4)$ and $Q = (x_1, y_1) = (3, 5)$, rounded to two decimal places is:

$$d = \{[3 - (-5)]^2 + [5 - (-4)]^2\}^{1/2}$$
$$= [(3 + 5)^2 + (5 + 4)^2]^{1/2}$$
$$= (8^2 + 9^2)^{1/2}$$
$$= (64 + 81)^{1/2}$$
$$= 145^{1/2}$$
$$= 12.04$$

PROBLEM 11-1
Plot the following points on the Cartesian plane: $(-2, 3)$, $(3, -1)$, $(0, 5)$, and $(-3, -3)$.

SOLUTION 11-1
These points are shown in Fig. 11-3. The dashed lines are perpendiculars, dropped to the axes to show the x and y coordinates of each point. (The dashed lines are not part of the coordinates themselves.)

PROBLEM 11-2
What is the distance between $(0, 5)$ and $(-3, -3)$ in Fig. 11-3? Express the answer to three decimal places.

SOLUTION 11-2
Use the distance formula. Let $(x_0, y_0) = (0, 5)$ and $(x_1, y_1) = (-3, -3)$. Then:

$$d = [(x_1 - x_0)^2 + (y_1 - y_0)^2]^{1/2}$$
$$= [(-3 - 0)^2 + (-3 - 5)^2]^{1/2}$$
$$= [(-3)^2 + (-8)^2]^{1/2}$$
$$= (9 + 64)^{1/2}$$
$$= 73^{1/2}$$
$$= 8.544$$

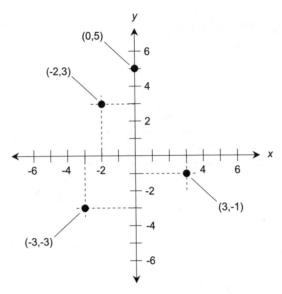

Fig. 11-3. Illustration for Problems 11-1 and 11-2.

Straight Lines in the Cartesian Plane

In Chapter 6, we saw what the graphs of quadratic equations look like in the Cartesian plane; they appear as parabolas. Here, we'll take a detailed look at how straight lines are plotted on the basis of linear equations in two variables. The standard form for this type of equation is as follows:

$$ax + by + c = 0$$

where a, b, and c are real-number constants, and the variables are x and y.

SLOPE–INTERCEPT FORM

A linear equation in variables x and y can be manipulated so it is in a form that is easy to plot on the Cartesian plane. Here is how a two-variable linear equation in standard form can be converted to *slope–intercept form*:

$$ax + by + c = 0$$
$$ax + by = -c$$
$$by = -ax - c$$
$$y = (-a/b)x - c/b$$
$$y = (-a/b)x + (-c/b)$$

where a, b, and c are real-number constants (the *coefficients* of the equation), and $b \neq 0$. The quantity $-a/b$ is called the *slope* of the line, an indicator of how steeply and in what sense the line slants. The quantity $-c/b$ represents the y value of the point at which the line crosses the y axis; this is called the *y-intercept*.

Let dx represent an arbitrarily small change in the value of x on such a graph. Let dy represent the tiny change in the value of y that results from this tiny change in x. The limit of the ratio dy/dx, as both dx and dy approach 0, is the slope of the line.

Suppose m is the slope of a line in Cartesian coordinates, and k is the y-intercept. The linear equation of that line can be written in slope–intercept form as:

$$y = (-a/b)x + (-c/b)$$

and therefore:

$$y = mx + k$$

To plot a graph of a linear equation in Cartesian coordinates using the slope–intercept form, proceed as follows:

- Convert the equation to slope–intercept form.
- Plot the point $x = 0, y = k$.
- Move to the right by n units on the graph.
- If m is positive, move upward mn units.
- If m is negative, move downward $|m|n$ units, where $|m|$ is the absolute value of m.
- If $m = 0$, don't move up or down at all.
- Plot the resulting point $x = n, y = mn + k$.
- Connect the two points with a straight line.

Figures 11-4A and B illustrate the following linear equations as graphed in slope–intercept form:

$$y = 5x - 3$$
$$y = -x + 2$$

A positive slope indicates that the line ramps upward as you move from left to right, and a negative slope indicates that the line ramps downward as you move from left to right. A slope of 0 indicates a horizontal line. The slope of a vertical line is undefined because, in the form shown here, it requires that m be defined as a quotient in which the denominator is equal to 0.

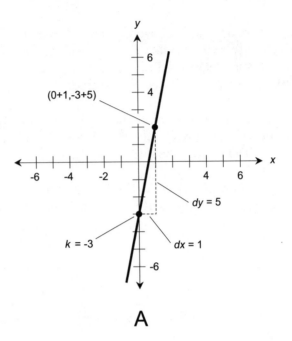

A

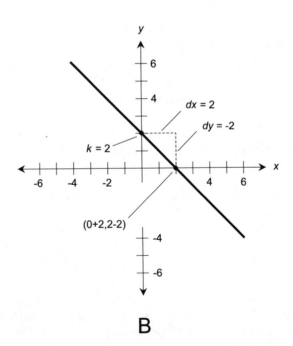

B

Fig. 11-4.   (A) Graph of the linear equation $y = 5x - 3$. (B) Graph of the linear equation $y = -x + 2$.

POINT–SLOPE FORM

It is difficult to plot a graph of a line based on the y-intercept (the point at which the line intersects the y axis) when the part of the graph of interest is far from the y axis. In this sort of situation, the *point–slope form* of a linear equation can be used. This form is based on the slope m of the line and the coordinates of a known point (x_0, y_0):

$$y - y_0 = m(x - x_0)$$

To plot a graph of a linear equation using the point–slope method, follow these steps in order:

- Convert the equation to point–slope form.
- Determine a point (x_0, y_0) by "plugging in" values.
- Plot (x_0, y_0) on the coordinate plane.
- Move to the right by n units on the graph, where n is some number that represents a reasonable distance on the graph.
- If m is positive, move upward mn units.
- If m is negative, move downward $|m|n$ units, where $|m|$ is the absolute value of m.
- If $m = 0$, don't move up or down at all.
- Plot the resulting point (x_1, y_1).
- Connect the points (x_0, y_0) and (x_1, y_1) with a straight line.

Figure 11-5A illustrates the following linear equation as graphed in point–slope form:

$$y - 104 = 3(x - 72)$$

Figure 11-5B is a graph of another linear equation in point–slope form:

$$y + 55 = -2(x + 85)$$

FINDING A LINEAR EQUATION BASED ON TWO POINTS

Suppose we are working in the Cartesian plane, and we know the exact coordinates of two points P and Q. These two points define a unique and distinct straight line. Call the line L. Let's give the coordinates of the points these names:

$$P = (x_p, y_p)$$
$$Q = (x_q, y_q)$$

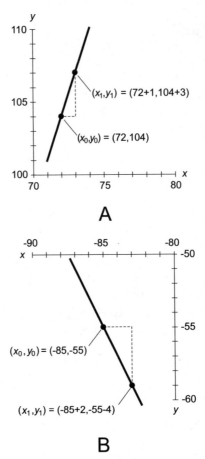

Fig. 11-5.   (A) Graph of the linear equation $y - 104 = 3(x - 72)$. (B) Graph of the linear equation $y + 55 = -2(x + 85)$.

The slope m of line L can be found using either of the following formulas:

$$m = (y_q - y_p)/(x_q - x_p)$$
$$m = (y_p - y_q)/(x_p - x_q)$$

These formulas work provided x_p is not equal to x_q. (If $x_p = x_q$, the denominators are equal to 0.) The point–slope equation of the line L can be determined based on the known coordinates of P or Q. Therefore, either of the following formulas represent the line L:

$$y - y_p = m(x - x_p)$$
$$y - y_q = m(x - x_q)$$

Reduced to standard form, these equations become:

$$mx - y + (y_p - mx_p) = 0$$
$$mx - y + (y_q - mx_q) = 0$$

That is, the coefficients a, b, and c are:

$$a = m$$
$$b = -1$$
$$c = y_p - mx_p = y_q - mx_q$$

FINDING A LINEAR EQUATION BASED ON y-INTERCEPT AND SLOPE

Suppose we are shown, on a graph, the coordinates of the point at which a straight line L crosses the y axis, and also the slope of that line. This information is sufficient to uniquely define the equation of L. Let's say that the y-intercept is equal to y_0 and the slope of the line is equal to m. Then the equation of the line is:

$$y - y_0 = mx$$

Reduced to standard form, this becomes:

$$mx - y + y_0 = 0$$

That is, the coefficients a, b, and c are:

$$a = m$$
$$b = -1$$
$$c = y_0$$

PROBLEM 11-3
Suppose you are shown Fig. 11-6. What is the equation of this line in standard form?

SOLUTION 11-3
The slope of the line is equal to 2; we are told this in the graph. Therefore, $m = 2$. The y-intercept is 4, so $y_0 = 4$. This means the equation of the line is:

$$2x - y + 4 = 0$$

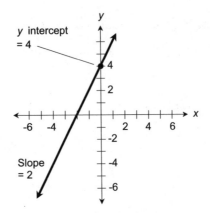

Fig. 11-6.   Illustration for Problem 11-3.

The Polar Coordinate Plane

Two versions of the polar coordinate plane are shown in Figs. 11-7 and 11-8. The independent variable is plotted as an angle θ (THAY-tuh) relative to a reference axis pointing to the right (or "east"), and the dependent variable is plotted as a distance (called the *radius*) r from the origin. (The origin is the center of the graph where $r = 0$.) A coordinate point is thus denoted in the form of an ordered pair (θ, r). In some texts, the independent and dependent variables are reversed, and the resulting ordered pairs are in the form (r, θ). But intuitively, it makes more sense to most people to plot the radius as a function of the angle, and not to plot the angle as a function of the radius. Here, we'll use the form (θ, r).

THE RADIUS

In any polar plane, the radii are shown by concentric circles. The larger the circle, the greater the value of r. In Figs. 11-7 and 11-8, the circles are not labeled in units. You can do that for yourself. Imagine each concentric circle, working outward, as increasing by any number of units you want. For example, each radial division might represent one unit, or five units, or 10, or 100.

THE DIRECTION

In polar coordinates, the direction can be expressed in angular degrees (symbolized °, where a full circle represents 360°) or in units called *radians*.

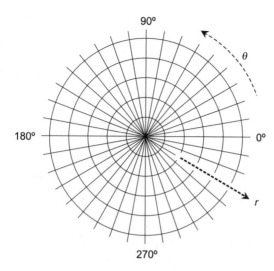

Fig. 11-7.   The polar coordinate plane. The angle θ is in degrees, and the radius r is in uniform increments.

What is a radian, you ask? Imagine two rays emanating outward from the center point of a circle. The rays each intersect the circle at a point. Call these points P and Q. Suppose the distance between P and Q, as measured along the arc of the circle, is equal to the radius of the circle. Then the measure of the angle between the rays is one radian (1 rad). There are 2π rad in a full circle, where π (the lowercase, non-italic Greek letter pi, pronounced "PIE") stands for the ratio of a circle's circumference to its diameter. This is a constant, and it happens to be an irrational number. The value of π is approximately 3.14159265359, often rounded off to 3.14159 or 3.14.

The direction in a polar coordinate system is usually defined counter-clockwise from a reference axis pointing to the right or "east." In Fig. 11-7, the direction θ is in degrees. Figure 11-8 shows the same polar plane, using radians to express the direction. (The "rad" abbreviation is not used, because it is obvious from the fact that the angles are multiples of π.) Regardless of whether degrees or radians are used, the angle on the graph is directly proportional to the value of θ.

NEGATIVE RADII

In polar coordinates, it is all right to have a negative radius. If some point is specified with $r < 0$, we multiply r by -1 so it becomes positive, and then add or subtract 180° (π rad) to or from the direction. That's like saying, "Go 3 kilometers (km) southeast" instead of "Go -3 km northwest." Negative

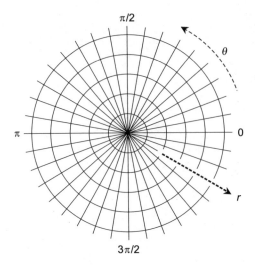

Fig. 11-8. Another form of the polar coordinate plane. The angle θ is in radians, and the radius r is in uniform increments.

radii must be allowed in order to graph figures that represent functions whose ranges can attain negative values.

NONSTANDARD DIRECTIONS

It's okay to have nonstandard direction angles in polar coordinates. If the value of θ is 360° (2π rad) or more, it represents more than one complete counterclockwise revolution from the 0° (0 rad) reference axis. If the direction angle is less than 0° (0 rad), it represents clockwise revolution instead of counterclockwise revolution. Nonstandard direction angles must be allowed in order to graph figures that represent functions whose domains go outside the standard angle range.

PROBLEM 11-4
Provide an example of a graphical object that can be represented as a function in polar coordinates, but not in Cartesian coordinates.

SOLUTION 11-4
Recall the definitions of the terms *relation* and *function*. When we talk about a function f, we can say that $r = f(\theta)$. A simple function of θ in polar coordinates is a *constant function* such as this:

$$f(\theta) = 3$$

Because $f(\theta)$ is just another way of denoting r, the radius, this function tells us that $r = 3$. This is a circle with a radius of 3 units.

In Cartesian coordinates, the equation of the circle with radius of 3 units is more complicated. It looks like this:

$$x^2 + y^2 = 9$$

(Note that $9 = 3^2$, the square of the radius.) If we let y be the dependent variable and x be the independent variable, we can rearrange the equation of the circle to get:

$$y = \pm(9 - x^2)^{1/2}$$

If we say that $y = g(x)$ where g is a function of x in this case, we are mistaken. There are values of x (the independent variable) that produce two values of y (the dependent variable). For example, when $x = 0$, $y = \pm 3$. If we want to say that g is a relation, that's fine, but we cannot call it a function.

Some Examples

In order to get a good idea of how the polar coordinate system works, let's look at the graphs of some familiar objects. Circles, ellipses, spirals, and other figures whose equations are complicated in Cartesian coordinates can often be expressed simply in polar coordinates. In general, the polar direction θ is expressed in radians. In the examples that follow, the "rad" abbreviation is eliminated, because it is understood that all angles are in radians.

CIRCLE CENTERED AT ORIGIN

The equation of a *circle* centered at the origin in the polar plane is given by the following formula:

$$r = a$$

where a is a real-number constant greater than 0. This is illustrated in Fig. 11-9.

CIRCLE PASSING THROUGH ORIGIN

The general form for the equation of a circle passing through the origin and centered at the point (θ_0, r_0) in the polar plane (Fig. 11-10) is as follows:

$$r = 2r_0 \cos (\theta - \theta_0)$$

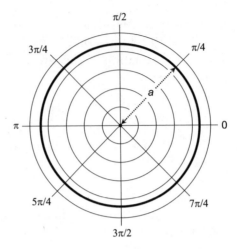

Fig. 11-9.   Polar graph of a circle centered at the origin.

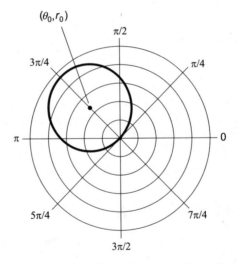

Fig. 11-10.   Polar graph of a circle passing through the origin.

ELLIPSE CENTERED AT ORIGIN

The equation of an *ellipse* centered at the origin in the polar plane is given by the following formula:

$$r = ab/(a^2 \sin^2 \theta + b^2 \cos^2 \theta)^{1/2}$$

where a and b are real-number constants greater than 0.

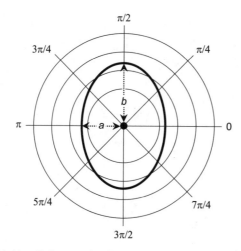

Fig. 11-11.　Polar graph of an ellipse centered at the origin.

Here, "sin" represents the sine function and "cos" represents the cosine function, both of which are available on any good scientific calculator. The exponent 2 attached to these functions indicates that after the sine or cosine has been found, the result should be squared.

In the ellipse, a represents the distance from the origin to the curve as measured along the "horizontal" ray $\theta = 0$, and b represents the distance from the origin to the curve as measured along the "vertical" ray $\theta = \pi/2$. This is illustrated in Fig. 11-11. The values a and b represent the lengths of the *semi-axes* of the ellipse; the greater value is the length of the *major semi-axis*, and the lesser value is the length of the *minor semi-axis*.

HYPERBOLA CENTERED AT ORIGIN

The general form of the equation of a *hyperbola* centered at the origin in the polar plane is given by the following formula:

$$r = ab/(b^2 \cos^2 \theta - a^2 \sin^2 \theta)^{1/2}$$

where a and b are real-number constants greater than 0.

Note how similar this is to the equation of the ellipse. But when we look at Fig. 11-12, which shows a hyperbola, it becomes apparent that the substitution of the minus sign for the plus sign in this equation makes a big difference!

Let D represent a rectangle whose center is at the origin, whose vertical edges are tangent to the hyperbola, and whose vertices (corners) lie on the

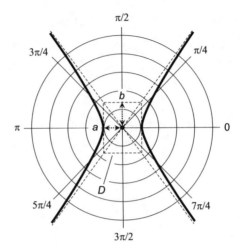

Fig. 11-12. Polar graph of a hyperbola centered at the origin.

asymptotes of the hyperbola (the long, straight, dashed lines in Fig. 11-12). Let *a* represent the distance from the origin to *D* as measured along the "horizontal" ray $\theta = 0$, and let *b* represent the distance from the origin to *D* as measured along the "vertical" ray $\theta = \pi/2$. The values *a* and *b* represent the lengths of the semi-axes of the hyperbola; the greater value is the length of the major semi-axis, and the lesser value is the length of the minor semi-axis.

LEMNISCATE

The general form of the equation of a *lemniscate* centered at the origin in the polar plane is given by the following formula:

$$r = a(\cos\ 2\theta)^{1/2}$$

where *a* is a real-number constant greater than 0. This is illustrated in Fig. 11-13.

THREE-LEAFED ROSE

The general form of the equation of a *three-leafed rose* centered at the origin in the polar plane is given by either of the following two formulas:

$$r = a\ \cos 3\theta$$

$$r = a\ \sin 3\theta$$

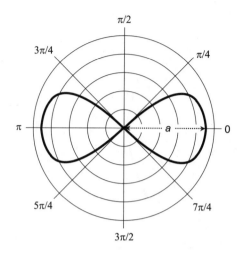

Fig. 11-13.   Polar graph of a lemniscate centered at the origin.

where a is a real-number constant greater than 0. The cosine version of the curve is illustrated in Fig. 11-14A. The sine version is illustrated in Fig. 11-14B.

FOUR-LEAFED ROSE

The general form of the equation of a *four-leafed rose* centered at the origin in the polar plane is given by either of the following two formulas:

$$r = a \cos 2\theta$$
$$r = a \sin 2\theta$$

where a is a real-number constant greater than 0. The cosine version is illustrated in Fig. 11-15A. The sine version is illustrated in Fig. 11-15B.

SPIRAL

The general form of the equation of a *spiral* centered at the origin in the polar plane is given by the following formula:

$$r = a\theta$$

where a is a real-number constant greater than 0. An example of this type of spiral, called the *spiral of Archimedes* because of the uniform manner in which its radius increases as the angle increases, is illustrated in Fig. 11-16.

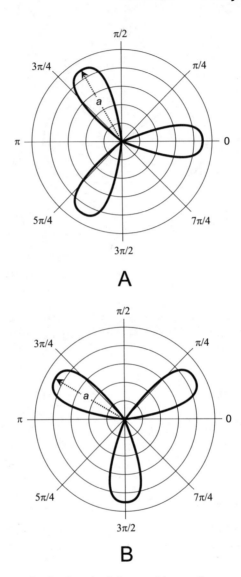

Fig. 11-14. (A) Polar graph of a three-leafed rose with equation $r = a \cos 3\theta$. (B) Polar graph of a three-leafed rose with equation $r = a \sin 3\theta$.

CARDIOID

The general form of the equation of a *cardioid* centered at the origin in the polar plane is given by the following formula:

$$r = 2a(1 + \cos\theta)$$

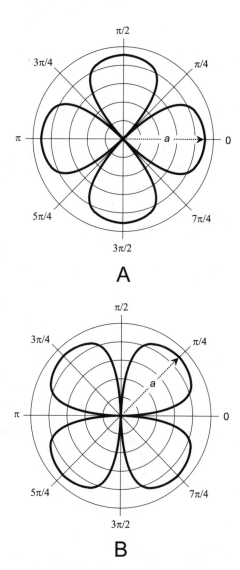

π/2
3π/4   π/4
π   a   0
5π/4   7π/4
3π/2
A

π/2
3π/4   π/4
π   a   0
5π/4   7π/4
3π/2
B

Fig. 11-15.   (A) Polar graph of a four-leafed rose with equation $r = a \cos 2\theta$. (B) Polar graph of a four-leafed rose with equation $r = a \sin 2\theta$.

where a is a real-number constant greater than 0. An example of this type of curve is illustrated in Fig. 11-17.

PROBLEM 11-5

What is the value of the constant, a, in the spiral shown in Fig. 11-16? What is the equation of this spiral? Assume that each radial division represents 1 unit.

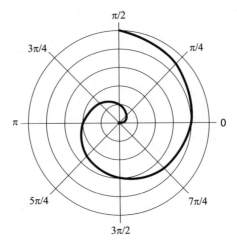

Fig. 11-16. Polar graph of a spiral; illustration for Problem 11-5.

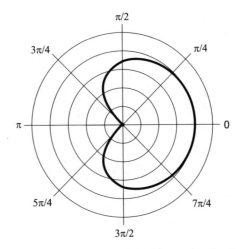

Fig. 11-17. Polar graph of a cardioid; illustration for Problem 11-6.

SOLUTION 11-5

Note that if $\theta = \pi$, then $r = 2$. Therefore, we can solve for a by substituting this number pair in the general equation for the spiral. We know that $(\theta, r) = (\pi, 2)$. Thus $2 = a\pi$. Solving for the constant, a, gives us $a = 2/\pi$. The equation of the spiral is:

$$r = (2/\pi)\theta$$

or, rearranged to get rid of the need for parentheses:

$$r = 2\theta/\pi$$

PROBLEM 11-6
What is the value of the constant, a, in the cardioid shown in Fig. 11-17?
What is the equation of this cardioid? Assume that each radial division
represents 1 unit.

SOLUTION 11-6
Note that if $\theta = 0$, then $r = 4$. We can solve for a by substituting this number
pair in the general equation for the cardioid. We know that $(\theta, r) = (0, 4)$.
Proceed like this:

$$r = 2a(1 + \cos\theta)$$
$$4 = 2a(1 + \cos 0)$$
$$4 = 2a(1 + 1)$$
$$4 = 4a$$
$$a = 1$$

This means that the equation of the cardioid is:

$$r = 2(1 + \cos\theta)$$

or, rearranged to get rid of the need for parentheses:

$$r = 2 + 2\,\cos\theta$$

Quiz

Refer to the text in this chapter if necessary. A good score is eight correct.
Answers are in the back of the book.

1. What is the distance between the points (3,4) and (–3,4) in Cartesian
   coordinates?
   (a) 10
   (b) 8
   (c) 6
   (d) 5

2. Suppose a straight line in Cartesian coordinates is represented by the
   following equation:

$$3x - 6y + 2 = 0$$

What is the slope of this line?
(a) 2
(b) 3/2
(c) 2/3
(d) 1/2

3. What is the distance between the origin and the point $(8, 8)$ in Cartesian coordinates?
   (a) 16
   (b) 8
   (c) $128^{1/2}$
   (d) $8^{1/2}$

4. Suppose the equation of a line in Cartesian coordinates is specified as:

$$y - 3 = 7(x + 2)$$

   From this equation, we know that the line passes through the point
   (a) $(2, 0)$
   (b) $(0, -3)$
   (c) $(2, 3)$
   (d) none of the above

5. What does the graph of $\theta = 0$ look like in polar coordinates?
   (a) A point.
   (b) A straight line.
   (c) A spiral.
   (d) A circle.

6. The term *ordinate* in Cartesian coordinates refers to
   (a) the direction
   (b) the value of the dependent variable
   (c) the value of the independent variable
   (d) the angle

7. A radian is equivalent to approximately
   (a) $114.6°$
   (b) $90°$
   (c) $57.3°$
   (d) $28.7°$

8. The direction of a point, relative to the origin, in conventional polar coordinates is expressed as an angle going
   (a) counterclockwise from the axis running from the origin upwards

(b) counterclockwise from the axis running from the origin towards the left
(c) counterclockwise from the axis running from the origin downwards
(d) counterclockwise from the axis running from the origin towards the right

9. What is the distance between the origin and $(r, \theta) = (4, 135°)$ in polar coordinates?
   (a) 0
   (b) 4
   (c) $3\pi/4$
   (d) It is impossible to determine this without more information.

10. A straight line has a slope of –3 in Cartesian coordinates, and runs through the point (0,2). The equation of this line is
   (a) $y = -3x + 2$
   (b) $y = 2x - 3$
   (c) $y - 3 = x + 2$
   (d) impossible to determine without more information

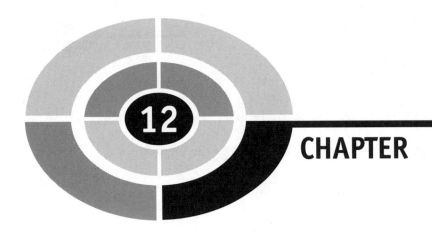

A Taste of Trigonometry

Trigonometry (or "trig") involves angles and distances. Trigonometry scares some people because of the Greek symbology, but the rules are clear-cut. Once you can get used to the idea of Greek letters representing angles, and if you're willing to draw diagrams and use a calculator, basic trigonometry loses most of its fear-inspiring qualities. You might even find yourself having fun with it.

More about Circles

Circles are defined by equations in which either x or y can be considered the dependent variable. We've already looked at the graphs of some simple circles in the xy-plane. Let's look more closely at this special species of geometric figure.

GENERAL EQUATION OF A CIRCLE

The equation in the xy-plane that represents a circle depends on two things: the radius of the circle, and the location of its center point.

Suppose r is the radius of a circle, expressed in arbitrary units. Imagine that the center point of the circle in Cartesian coordinates is located at the point $x = a$ and $y = b$, represented by the ordered pair (a, b). Then the equation of that circle is:

$$(x - a)^2 + (y - b)^2 = r^2$$

If the center of the circle happens to be at the origin, that is, at $(0, 0)$ on the coordinate plane, then the general equation is simpler:

$$x^2 + y^2 = r^2$$

THE UNIT CIRCLE

Consider a circle in rectangular coordinates with the following equation:

$$x^2 + y^2 = 1$$

This is called the *unit circle* because its radius is 1 unit, and it is centered at the origin $(0, 0)$. This circle gives us a good way to define the common trigonometric functions, which are sometimes called *circular functions*.

IT'S GREEK TO US

In trigonometry, mathematicians and scientists have acquired the habit of using Greek letters to represent angles. The most common symbol for this purpose is an italic, lowercase Greek theta (pronounced "THAY-tuh"). It looks like an italic numeral 0 with a horizontal line through it (θ).

When writing about two different angles, a second Greek letter is used along with θ. Most often, it is the italic, lowercase letter phi (pronounced "FIE" or "FEE"). This character looks like an italic lowercase English letter o with a forward slash through it (ϕ).

Sometimes the italic, lowercase Greek alpha ("AL-fuh"), beta ("BAY-tuh"), and gamma ("GAM-uh") are used to represent angles. These, respectively, look like this: α, β, γ. When things get messy and there are a lot of angles to talk about, numeric subscripts are sometimes used with Greek letters, so don't be surprised if you see text in which angles are denoted θ_1, θ_2, θ_3, and so on.

ANGULAR UNITS

There are two main units by which the measures of angles in a flat plane can be specified: the radian and the degree. The radian was defined in the previous chapter. A quarter circle is $\pi/2$ rad, a half circle is π rad, and three quarters of a circle is $3\pi/2$ rad. Mathematicians prefer to use the radian when working with trigonometric functions, and the abbreviation "rad" is left out. So if you see something like $\theta_1 = \pi/4$, you know the angle θ_1 is expressed in radians.

The angular degree (°), also called the *degree of arc*, is the unit of angular measure most familiar to lay people. One degree (1°) is $1/360$ of a full circle. An angle of 90° represents a quarter circle, 180° represents a half circle, 270° represents three quarters of a circle, and 360° represents a full circle. A right angle has a measure of 90°, an acute angle has a measure of more than 0° but less than 90°, and an obtuse angle has a measure of more than 90° but less than 180°.

To denote the measures of tiny angles, or to precisely denote the measures of angles in general, smaller units are used. One *minute of arc* or *arc minute*, symbolized by an apostrophe or accent (′) or abbreviated as m or min, is $1/60$ of a degree. One *second of arc* or *arc second*, symbolized by a closing quotation mark (″) or abbreviated as s or sec, is $1/60$ of an arc minute or $1/3600$ of a degree. An example of an angle in this notation is $30°\,15'\,0''$, which is read as "30 degrees, 15 minutes, 0 seconds."

Alternatively, fractions of a degree can be denoted in decimal form. You might see, for example, 30.25°. This is the same as $30°\,15'\,0''$. Decimal fractions of degrees are easier to work with than the minute/second scheme when angles must be added and subtracted, or when using a conventional calculator to work out trigonometry problems. Nevertheless, the minute/second system, like the English system of measurements, remains in widespread use.

PROBLEM 12-1
A text discussion tells you that $\theta_1 = \pi/4$. What is the measure of θ_1 in degrees?

SOLUTION 12-1
There are 2π rad in a full circle of 360°. The value $\pi/4$ is equal to $1/8$ of 2π. Therefore, the angle θ_1 is $1/8$ of a full circle, or 45°.

PROBLEM 12-2
Suppose your town is listed in an almanac as being at $40°\,20'$ north latitude and $93°\,48'$ west longitude. What are these values in decimal form? Express your answers to two decimal places.

SOLUTION 12-2
There are 60 minutes of arc in one degree. To calculate the latitude, note that
$20' = (20/60)° = 0.33°$; that means the latitude is 40.33° north. To calculate
the longitude, note that $48' = (48/60)° = 0.80°$; that means the longitude is
93.80° west.

Primary Circular Functions

Consider a circle in the Cartesian xy-plane with the following equation:

$$x^2 + y^2 = 1$$

This equation, as defined earlier in this chapter, represents the unit circle.
Let θ be an angle whose apex is at the origin, and that is measured counterclockwise from the x axis, as shown in Fig. 12-1. Suppose this angle corresponds to a ray that intersects the unit circle at some point $P = (x_0, y_0)$. We
can define the three basic circular functions, also called the *primary
circular functions*, of the angle θ in an elegant way. But before we get into
this, let's extend our notion of angles to cover negative values, and also to
cover values more than 360° (2π rad).

OFFBEAT ANGLES

In trigonometry, any *direction angle* expressible as a real number, no matter
how extreme, can always be reduced to something that is at least 0° (0 rad)

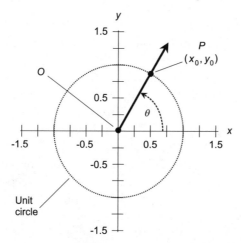

Fig. 12-1.   The unit circle is the basis for the trigonometric functions.

but less than 360° (2π rad). If you look at Fig. 12-1, you should be able to envision how this works. Even if the ray OP makes more than one complete revolution counterclockwise from the x axis, or if it turns clockwise instead, its direction can always be defined by some counterclockwise angle of least 0° (0 rad) but less than 360° (2π rad) relative to the x axis.

Any offbeat direction angle such as 730° or $-9\pi/4$ rad can be reduced to a direction angle that measures at least 0° (0 rad) but less than 360° (2π rad) by adding or subtracting some whole-number multiple of 360° (2π rad). But you have to be careful about this. A direction angle specifies orientation only. The orientation of the ray OP is the same for an angle of 540° (3π rad) as for an angle of 180° (π rad), but the larger value carries with it the insinuation that the ray (also called a *vector*) OP has revolved 1.5 times around, while the smaller angle implies that it has undergone less than one complete revolution. Sometimes this doesn't matter, but often it does!

Negative angles are encountered in trigonometry, especially in graphs of functions. Multiple revolutions of objects are important in physics and engineering. So if you ever hear or read about an angle such as −90°, −π/2 rad, 900°, or 5π rad, you can be confident that it has meaning. The negative value indicates clockwise rotation. An angle that is said to measure more than 360° (2π rad) indicates more than one complete revolution counterclockwise. An angle that is said to measure less than −360° (−2π rad) indicates more than one revolution clockwise.

THE SINE FUNCTION

In Fig. 12-1, imagine ray OP pointing along the x axis, and then starting to rotate counterclockwise on its end point O, as if point O is a mechanical bearing. The point P, represented by coordinates (x_0, y_0), thus revolves around point O, following the perimeter of the unit circle.

Imagine what happens to the value of y_0 (the ordinate of point P) during one complete revolution of ray OP. The ordinate of P starts out at $y_0 = 0$, then increases until it reaches $y_0 = 1$ after P has gone 90° or π/2 rad around the circle ($\theta = 90° = \pi/2$). After that, y_0 begins to decrease, getting back to $y_0 = 0$ when P has gone 180° or π rad around the circle ($\theta = 180° = \pi$). As P continues on its counterclockwise trek, y_0 keeps decreasing until, at $\theta = 270° = 3\pi/2$, the value of y_0 reaches its minimum of −1. After that, the value of y_0 rises again until, when P has gone completely around the circle, it returns to $y_0 = 0$ for $\theta = 360° = 2\pi$.

The value of y_0 is defined as the *sine* of the angle θ. The *sine function* is abbreviated sin, so we can state this simple equation:

$$\sin \theta = y_0$$

CIRCULAR MOTION

Imagine that you attach a "glow-in-the-dark" ball to the end of a string, and then swing the ball around and around at a rate of one revolution per second. The ball describes a circle as viewed from high above, as shown in Fig. 12-2A. Suppose you make the ball circle your head so the path of the ball lies in a perfectly horizontal plane. Imagine that you are in the middle of a flat, open field on a perfectly dark night. If a friend stands far away with his or her eyes exactly in the plane of the ball's orbit and looks at you and the ball through a pair of binoculars, what does your friend see?

Your friend, watching from a great distance and from a viewpoint exactly in the plane defined by the ball's orbit, sees a point of light that oscillates back and forth, from right-to-left and left-to-right, along what appears to be a straight-line path (Fig. 12-2B). Starting from its extreme right-most apparent position, the glowing point moves toward the left for half a second, speeding up and then slowing down; then it stops and reverses direction; then it moves toward the right for half a second, speeding up and then slowing down; then it stops and turns around again. This goes on and on for as long as you care to swing the ball around your head. As seen by your friend, the ball reaches its extreme right-most position at one-second intervals because its orbital speed is one revolution per second.

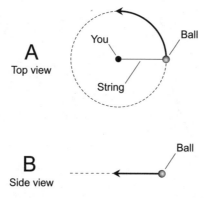

Fig. 12-2.   Orbiting ball and string. At A, as seen from above; at B, as seen edge-on.

THE SINE WAVE

If you graph the apparent position of the ball as seen by your friend with respect to time, the result is a *sine wave*, which is a graphical plot of a sine function. Some sine waves "rise higher and lower" (corresponding to a longer string), some are "flatter" (a shorter string), some are "stretched out" (a slower rate of revolution), and some are "squashed" (a faster rate of revolution). But the characteristic shape of the wave, known as a *sinusoid*, is the same in every case.

You can whirl the ball around faster or slower than one revolution per second. The string can be made longer or shorter. These adjustments alter the *frequency* and/or the *amplitude* of the sine wave. But any sinusoid can be defined as the path of a point that orbits a central point in a perfect circle, with a certain radius, at a certain rate of revolution.

Circular motion in the Cartesian plane can be defined in terms of a general formula:

$$y = a \sin b\theta$$

where a is a constant that depends on the radius of the circle, and b is a constant that depends on the revolution rate. The angle θ is expressed in a counterclockwise direction from the positive x axis. Figure 12-3 illustrates a graph of the basic sine function: a sine wave for which $a=1$ and $b=1$, and for which the angle is expressed in radians.

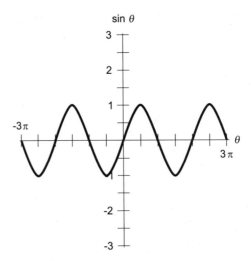

Fig. 12-3.   Graph of the sine function for values of θ between -3π rad and 3π rad.

THE COSINE FUNCTION

Look again at Fig. 12-1. Imagine, once again, a ray from the origin outward through point P on the circle, pointing at first along the x axis, and then rotating in a counterclockwise direction.

What happens to the value of x_0 during one complete revolution of the ray? It starts out at $x_0 = 1$, then decreases until it reaches $x_0 = 0$ when $\theta = 90° = \pi/2$. After that, x_0 continues to decrease, getting down to $x_0 = -1$ when $\theta = 180° = \pi$. As P continues counterclockwise around the circle, x_0 begins to increase again; at $\theta = 270° = 3\pi/2$, the value gets back up to $x_0 = 0$. After that, x_0 increases further until, when P has gone completely around the circle, it returns to $x_0 = 1$ for $\theta = 360° = 2\pi$.

The value of x_0 is defined as the *cosine* of the angle θ. The *cosine function* is abbreviated cos. So we can write this:

$$\cos \theta = x_0$$

THE COSINE WAVE

Circular motion in the Cartesian plane can be defined in terms of the cosine function as well as the sine function:

$$y = a \cos b\theta$$

where a is a constant that depends on the radius of the circle, and b is a constant that depends on the revolution rate, just as is the case with the sine function. The angle θ is measured or defined counterclockwise from the positive x axis, also after the fashion of the sine function.

The shape of a cosine wave is exactly the same as the shape of a sine wave. Both waves are sinusoids. But the cosine wave is shifted in phase by 1/4 cycle (90° or $\pi/2$ rad) with respect to the sine wave. Figure 12-4 illustrates a graph of the basic cosine function: a cosine wave for which $a = 1$ and $b = 1$, and for which the angle is expressed in radians.

THE TANGENT FUNCTION

Once again, refer to Fig. 12-1. The *tangent* (abbreviated tan) of an angle θ is defined using the same ray OP and the same point $P = (x_0, y_0)$ as is done with the sine and cosine functions. The definition is:

$$\tan \theta = y_0/x_0$$

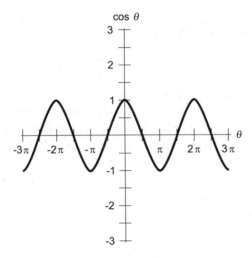

Fig. 12-4.   Graph of the cosine function for values of θ between -3π rad and 3π rad.

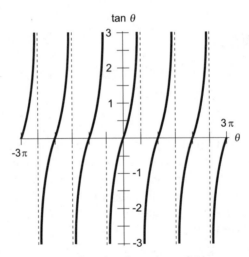

Fig. 12-5.   Graph of the tangent function for values of θ between -3π rad and 3π rad.

We already know that $\sin \theta = y_0$ and $\cos \theta = x_0$, so we can express the tangent function in terms of the sine and the cosine:

$$\tan \theta = \sin \theta / \cos \theta$$

This function is interesting because, unlike the sine and cosine functions, it "blows up" at certain values of θ. This is shown by the graph of the tangent function (Fig. 12-5). Whenever $x_0 = 0$, the denominator of either quotient above becomes zero. Division by zero is not defined, and that means the tangent function is not defined for any angle θ such that $\cos \theta = 0$. Such angles are all the positive and negative odd-integer multiples of 90° ($\pi/2$ rad).

PROBLEM 12-3

What is tan 45°? Do not perform any calculations. You should be able to infer this without having to write down a single numeral.

SOLUTION 12-3

Draw a diagram of a unit circle, such as the one in Fig. 12-1, and place ray OP such that it subtends an angle of 45° with respect to the x axis. That angle is the angle for which we want to find the tangent. Note that the ray OP also subtends an angle of 45° with respect to the y axis, because the x and y axes are perpendicular (they are oriented at 90° with respect to each other), and 45° is exactly half of 90°. Every point on the ray OP is equally distant from the x and y axes; this includes the point (x_0, y_0). It follows that $x_0 = y_0$, and neither of them is equal to zero. From this, we can conclude that $y_0/x_0 = 1$. According to the definition of the tangent function, therefore, $\tan 45° = 1$.

Secondary Circular Functions

The three functions defined above form the cornerstone for trigonometry. But three more circular functions exist. Their values represent the reciprocals of the values of the primary circular functions.

THE COSECANT FUNCTION

Imagine the ray OP in Fig. 12-1, at an angle θ with respect to the x axis, pointing outward from the origin, and intersecting the unit circle at the point $P = (x_0, y_0)$. The reciprocal of the ordinate, that is, $1/y_0$, is defined as the *cosecant* of the angle θ. The cosecant function is abbreviated csc, so we can say this in mathematical terms:

$$\csc \theta = 1/y_0$$

The cosecant function is the reciprocal of the sine function. For any angle θ, the following equation is always true as long as sin θ is

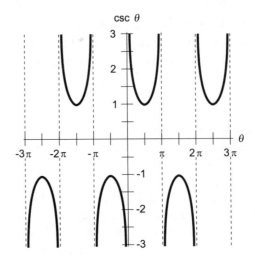

Fig. 12-6.   Graph of the cosecant function for values of θ between -3π rad and 3π rad.

not equal to zero:

$$\csc\theta = 1/\sin\theta$$

The cosecant function is not defined for $0°$ (0 rad), or for any multiple of $180°$ (π rad). This is because the sine of any such angle is equal to 0, which would mean that the cosecant would have to be $1/0$. We can't do anything with a quotient in which the denominator is 0.

Figure 12-6 is a graph of the cosecant function for values of θ between -3π and 3π. Note the angles at which the function "explodes."

THE SECANT FUNCTION

Now consider the reciprocal of the abscissa, that is, $1/x_0$, in Fig. 12-1. This is the *secant* of the angle θ. The secant function is abbreviated sec, so we can define it like this:

$$\sec\theta = 1/x_0$$

The secant of any angle is the reciprocal of the cosine of that angle. As long as $\cos\theta$ is not equal to zero, the following is true:

$$\sec\theta = 1/\cos\theta$$

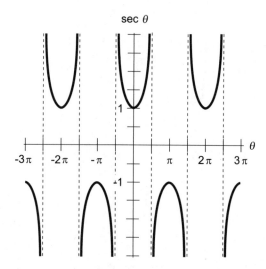

Fig. 12-7.   Graph of the secant function for values of θ between -3π rad and 3π rad.

The secant function is not defined for $90°$ ($\pi/2$ rad), or for any positive or negative odd multiple thereof. Figure 12-7 is a graph of the secant function for values of θ between -3π and 3π. Note the angles at which the function "blows up."

THE COTANGENT FUNCTION

There's one more circular function. Consider the value of x_0/y_0 at the point P where the ray OP crosses the unit circle. This quotient is called the *cotangent* of the angle θ. The word "cotangent" is abbreviated as cot. For any ray anchored at the origin and crossing the unit circle at an angle θ:

$$\cot\theta = x_0/y_0$$

Because we already know that $\sin\theta = y_0$ and $\cos\theta = x_0$, we can express the cotangent function in terms of the sine and the cosine:

$$\cot\theta = \cos\theta/\sin\theta$$

The cotangent function is also the reciprocal of the tangent function:

$$\cot\theta = 1/\tan\theta$$

This function, like the tangent function, "explodes" at certain values of θ. Whenever $y_0 = 0$, the denominator of either quotient above becomes zero, and the cotangent function is not defined. This occurs at all integer multiples

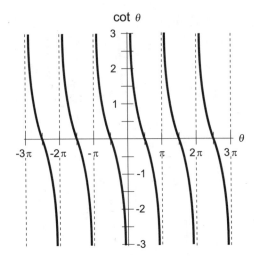

Fig. 12-8.   Graph of the cotangent function for values of θ between -3π rad and 3π rad.

of 180° (π rad). Figure 12-8 is a graph of the cotangent function for values of θ between -3π and 3π.

VALUES OF CIRCULAR FUNCTIONS

Now that you know how the circular functions are defined, you might wonder how the values are calculated. The answer: with a calculator! Most personal computers have a calculator program built into the operating system. You might have to dig around in the folders to find it, but once you do, you can put a shortcut to it on your computer's desktop. Put the calculator in the "scientific" mode.

The values of the sine and cosine function never get smaller than -1 or larger than 1. The values of other functions can vary wildly. Put a few numbers into your calculator and see what happens when you apply the circular functions to them. Pay attention to whether you're using degrees or radians. When the value of a function "blows up" (the denominator in the unit-circle equation defining it becomes zero), you'll get an error message on the calculator.

PROBLEM 12-4

Use a portable scientific calculator, or the calculator program in a personal computer, to find the values of all six circular functions of 66°. Round your answers off to three decimal places. If your calculator does not have keys for the cosecant (csc), secant (sec), or cotangent (cot) functions, first find the sine (sin), cosine (cos), and tangent (tan) respectively, then find the reciprocal, and finally round off your answer to three decimal places.

SOLUTION 12-4
You should get the following results. Be sure your calculator is set to work with degrees, not radians.

$$\sin 66° = 0.914$$

$$\cos 66° = 0.407$$

$$\tan 66° = 2.246$$

$$\csc 66° = 1/(\sin 66°) = 1.095$$

$$\sec 66° = 1/(\cos 66°) = 2.459$$

$$\cot 66 = 1/(\tan 66°) = 0.445$$

The Right Triangle Model

We have just defined the six circular functions – sine, cosine, tangent, cosecant, secant, and cotangent – in terms of points on a circle. There is another way to define these functions: the *right-triangle model*.

TRIANGLE AND ANGLE NOTATION

In geometry, it is customary to denote triangles by writing an uppercase Greek letter delta (Δ) followed by the names of the three points representing the corners, or *vertices*, of the triangle. For example, if P, Q, and R are the names of three points, then $\triangle PQR$ is the triangle formed by connecting these points with straight line segments. We read this as "triangle PQR."

Angles are denoted by writing the symbol ∠ (which resembles an extremely italicized, uppercase English letter L without serifs) followed by the names of three points that uniquely determine the angle. This scheme lets us specify the extent and position of the angle, and also the rotational sense in which it is expressed. For example, if there are three points P, Q, and R, then $\angle PQR$ (read "angle PQR") has the same measure as $\angle RQP$, but in the opposite direction. The middle point, Q, is the *vertex* of the angle.

The rotational sense in which an angle is measured can be significant in physics, astronomy, and engineering, and also when working in coordinate systems. In the Cartesian plane, remember that angles measured counterclockwise are considered positive, while angles measured clockwise are considered negative. If we have $\angle PQR$ that measures $30°$ around a circle in Cartesian coordinates, then $\angle RQP$ measures $-30°$, whose direction is equivalent to an angle of $330°$. The cosines of these two angles happen to be the same, but the sines differ.

RATIOS OF SIDES

Consider a right triangle defined by points P, Q, and R, as shown in Fig. 12-9. Suppose that $\angle QPR$ is a right angle, so $\triangle PQR$ is a *right triangle*. Let d be the length of line segment QP, e be the length of line segment PR, and f be the length of line segment QR. Let θ be $\angle PQR$, the angle measured counterclockwise between line segments QP and QR. The six circular trigonometric functions can be defined as ratios between the lengths of the sides, as follows:

$$\sin \theta = e/f$$
$$\cos \theta = d/f$$
$$\tan \theta = e/d$$
$$\csc \theta = f/e$$
$$\sec \theta = f/d$$
$$\cot \theta = d/e$$

The longest side of a right triangle is always opposite the $90°$ angle, and is called the *hypotenuse*. In Fig. 12-9, this is the side QR whose length is f. The other two sides are called *adjacent sides* because they are both adjacent to the right angle.

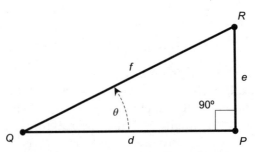

Fig. 12-9. The right-triangle model for defining trigonometric functions. All right triangles obey the theorem of Pythagoras. Illustration for Problems 12-5 and 12-6.

SUM OF ANGLE MEASURES

The following fact can be useful in deducing the measures of angles in trigonometric calculations. It's a simple theorem in geometry that you should remember. In any triangle, the sum of the measures of the interior angles is 180° (π rad). This is true whether it is a right triangle or not, as long as all the angles are measured in the plane defined by the three vertices of the triangle.

RANGE OF ANGLES

In the right-triangle model, the values of the circular functions are defined only for angles between, but not including, 0° and 90° (0 rad and $\pi/2$ rad). All angles outside this range are better dealt with using the unit-circle model.

Using the right-triangle scheme, a trigonometric function is undefined whenever the denominator in its "side ratio" (according to the formulas above) is equal to zero. The length of the hypotenuse (side f) is never zero, but if a right triangle is "squashed" or "squeezed" flat either horizontally or vertically, then the length of one of the adjacent sides (d or e) can become zero. Such objects aren't triangles in the strict sense, because they have only two vertices rather than three.

PROBLEM 12-5

Suppose there is a triangle whose sides are 3, 4, and 5 units, respectively. What is the sine of the angle θ opposite the side that measures 3 units? Express your answer to three decimal places.

SOLUTION 12-5

If we are to use the right-triangle model to solve this problem, we must first be certain that a triangle with sides of 3, 4, and 5 units is a right triangle. Otherwise, the scheme won't work. We can test for this by seeing if the Pythagorean theorem applies. If this triangle is a right triangle, then the side measuring 5 units is the hypotenuse, and we should find that $3^2 + 4^2 = 5^2$. Checking, we see that $3^2 = 9$ and $4^2 = 16$. Therefore, $3^2 + 4^2 = 9 + 16 = 25$, which is equal to 5^2. It's a right triangle, indeed!

It helps to draw a picture here, after the fashion of Fig. 12-9. Put the angle θ, which we are analyzing, at lower left (corresponding to the vertex point Q). Label the hypotenuse $f = 5$. Now we must figure out which of the other sides should be called d, and which should be called e. We want to find the sine of the angle opposite the side whose length is 3 units, and this angle, in Fig. 12-9, is opposite side PR, whose length is equal to e. So we set $e = 3$. That leaves us with no other choice for d than to set $d = 4$.

According to the formulas above, the sine of the angle in question is equal to e/f. In this case, that means $\sin\theta = 3/5 = 0.600$.

PROBLEM 12-6
What are the values of the other five circular functions for the angle θ as defined in Problem 12-5? Express your answers to three decimal places.

SOLUTION 12-6
Plug numbers into the formulas given above, representing the ratios of the lengths of sides in the right triangle:

$$\cos\theta = d/f = 4/5 = 0.800$$
$$\tan\theta = e/d = 3/4 = 0.750$$
$$\csc\theta = f/e = 5/3 = 1.667$$
$$\sec\theta = f/d = 5/4 = 1.250$$
$$\cot\theta = d/e = 4/3 = 1.333$$

Pythagorean Extras

The theorem of Pythagoras, which you learned earlier in this book, can be extended to cover two important facts involving the circular trigonometric functions. These are worth remembering.

PYTHAGOREAN THEOREM FOR SINE AND COSINE

The sum of the squares of the sine and cosine of an angle is always equal to 1. In mathematical terms, we can write it like this:

$$(\sin\theta)^2 + (\cos\theta)^2 = 1$$

When the value of a trigonometric function is squared, the exponent 2 is customarily placed after the abbreviation of the function, so the parentheses can be eliminated from the expression. Therefore, the above equation is more often written this way:

$$\sin^2\theta + \cos^2\theta = 1$$

PYTHAGOREAN THEOREM FOR SECANT AND TANGENT

The difference between the squares of the secant and tangent of an angle is always equal to either 1 or −1. The following formulas apply for all angles except $\theta = 90°$ ($\pi/2$ rad) and $\theta = 270°$ ($3\pi/2$ rad):

$$\sec^2 \theta - \tan^2 \theta = 1$$
$$\tan^2 \theta - \sec^2 \theta = -1$$

USE CRUTCHES!

Trigonometry is a branch of mathematics with extensive applications in science, engineering, architecture – even in art. You should not be shy about using a calculator and making sketches to help yourself solve problems. So what if they're "crutches"? As long as you get the correct answer, you can use any help you need. (Just be sure you keep a fresh, spare battery around for your calculator if it's battery powered.)

PROBLEM 12-7

Use a drawing of the unit circle to help show why it is true that $\sin^2 \theta + \cos^2 \theta = 1$ for angles θ greater than $0°$ and less than $90°$. (Hint: a right triangle is involved.)

SOLUTION 12-7

Figure 12-10 shows a drawing of the unit circle, with θ defined counterclockwise between the x axis and a ray emanating from the origin. When the angle is greater than $0°$ but less than $90°$, a right triangle is formed, with a segment of the ray as the hypotenuse. The length of this segment is equal to the radius of the unit circle. This radius, by definition, is 1 unit. According to the Pythagorean theorem for right triangles, the square of the length of the hypotenuse is equal to the sum of the squares of the lengths of the other

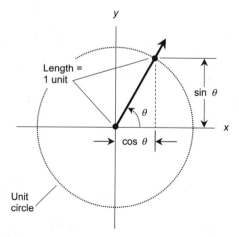

Fig. 12-10.   Illustration for Problem 12-7.

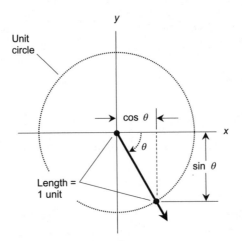

Fig. 12-11.   Illustration for Problem 12-8.

two sides. It is easy to see from Fig. 12-9 that the lengths of these other two sides are $\sin\theta$ and $\cos\theta$. Therefore:

$$(\sin\theta)^2 + (\sin\theta)^2 = 1^2$$

which is the same as saying that $\sin^2\theta + \cos^2\theta = 1$.

PROBLEM 12-8
Use another drawing of the unit circle to help show why it is true that $\sin^2\theta + \cos^2\theta = 1$ for angles θ greater than $270°$ and less than $360°$. (Hint: this range can be considered, in directional terms, as equivalent to the range of angles greater than $-90°$ and less than $0°$.)

SOLUTION 12-8
Figure 12-11 shows how this can be done. Draw a mirror image of Fig. 12-10, with the angle θ defined clockwise instead of counterclockwise. Again we have a right triangle. This triangle, like all right triangles, obeys the Pythagorean theorem.

Quiz

Refer to the text in this chapter if necessary. A good score is eight correct. Answers are in the back of the book.

1. Suppose you stand on a flat, level surface on a sunny morning. Your shadow is exactly four times as long as you are tall. What is the angle

of the sun above the horizon, to the nearest degree?
(a) 14°
(b) 45°
(c) 76°
(d) More information is needed to answer this question.

2. Suppose you stand on a flat, level surface on a sunny morning. The sun is
    at an angle of 50° above the horizon. You are exactly 160 centimeters
    (cm) tall. How long is your shadow, to the nearest centimeter?
    (a) 122 cm
    (b) 134 cm
    (c) 190 cm
    (d) More information is needed to answer this question.

3. Imagine that you stand some distance away from a tall building with a
    flat roof. The earth in the vicinity of the building is flat and level. The
    angle between the horizon and the top of the building, as measured
    from a point at your feet, is 70.00°. The straight-line distance between
    your feet and the nearest point at the top of the building is 200 meters
    (m). How tall is the building, to the nearest meter?
    (a) 115 m
    (b) 188 m
    (c) 213 m
    (d) More information is needed to answer this question.

4. One arc second is the equivalent of
    (a) 1/60 of an arc minute
    (b) 1/3600 of an arc minute
    (c) $1/\pi$ of an arc minute
    (d) $1/(2\pi)$ of an arc minute

5. Approximately how many degrees are there in 1 rad?
    (a) 32.7°
    (b) 45°
    (c) 57.3°
    (d) None of the above.

6. Suppose an almanac lists the latitude of your town as 35° 15′ 00″ N.
    This is equivalent to
    (a) 35.15° N
    (b) 35.25° N
    (c) 35.50° N
    (d) None of the above.

7. What is the angle, in radians, representing revolution by 3/4 of a circle in a clockwise direction?
    (a) $3\pi/4$
    (b) $-3\pi/4$
    (c) $3\pi/2$
    (d) $-3\pi/2$

8. The circumference of the earth is approximately 40,000 kilometers (km). What is the angular separation between two points 111 km apart, measured over the shortest possible path on the surface? Neglect irregularities in the terrain. Assume the earth is a perfect sphere.
    (a) 1 rad
    (b) 57.3°
    (c) 1/360 rad
    (d) 1°

9. Suppose you live in a remote area and decide to install an outdoor antenna for receiving stereo FM broadcasts. You buy a support tower that is 20.00 meters (m) high. The installation guide recommends guying the tower from a point halfway up, and also from the top. The guide also recommends that all guy wires be sloped at angles no greater than 45° with respect to the horizontal. Suppose you install two sets of guy wires, one set from a point halfway up and the other set from the top. All guys are fastened to three anchors at ground level. Each guy anchor is at just the right distance from the base of the tower to make the top set of guy wires slant at precisely 45.00°. At what angle, to the nearest tenth of a degree and relative to the horizontal, does each of the lower-level guy wires slant? Assume the ground in the vicinity of the tower is flat and level.
    (a) 22.5°
    (b) 26.6°
    (c) 30.0°
    (d) More information is needed to answer this question.

10. What is the largest possible value for the tangent of an angle?
    (a) 1
    (b) 90°
    (c) π rad
    (d) None of the above.

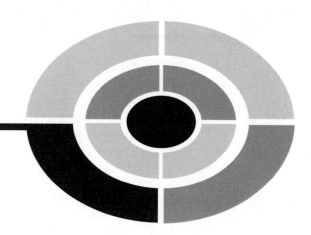

Test: Part 3

Do not refer to the text when taking this test. You may draw diagrams or use a calculator if necessary. A good score is at least 30 correct. Answers are in the back of the book. It's best to have a friend check your score the first time, so you won't memorize the answers if you want to take the test again.

1. Imagine that you are on a long, straight, level stretch of two-lane highway. A set of utility poles, all 20 m high, 40 m to the right of the center line (as you see them while driving), and 70 m from each other, is set up alongside the road. A perfectly taut, straight wire runs along the tops of the poles. Which of the following statements (a), (b), (c), or (d), if any, is true?

(a) The line of the wire is parallel to the highway center line.

(b) The line of the wire is skewed with respect to the highway center line.

(c) If the line of the wire and the highway center line were extended indefinitely, they would intersect at some point far in front of you.

(d) If the line of the wire and the highway center line were extended indefinitely, they would intersect at some point far behind you.

(e) None of the above statements (a), (b), (c), or (d) is true.

2. In a right triangle, the hypotenuse
   (a) is always opposite the smallest angle
   (b) is always opposite the 30° angle
   (c) is always opposite the 60° angle
   (d) is always opposite the 120° angle
   (e) is always opposite the largest angle

3. Suppose you are an ant on a flat, level field, and you are looking at a broadcast tower. Your companion, another ant, looks up at the tower and says, "We are only 0.001 m tall, and that building is 100,000 times as tall as we are." To this you reply, "The top of that tower is at an angle of about 45° with respect to the horizon." Your companion, not to be outwitted, correctly observes:
   (a) "If that is true, then we are roughly 10 m from the bottom of the tower."
   (b) "If that is true, then we are roughly 71 m from the bottom of the tower."
   (c) "If that is true, then we are roughly 100 m from the bottom of the tower."
   (d) "If that is true, then we are roughly 710 m from the bottom of the tower."
   (e) "That tells us nothing about our distance from the tower."

4. Imagine a city laid out in square blocks on a level expanse of prairie, with "streets" running east–west and "avenues" running north–south. The distances between adjacent intersections are all exactly 200 m. You stop at Gas Station Alpha and fuel up the car, and then you ask the attendant how to get to Junk Market Beta. The attendant says, "Go 5 blocks north, then turn left and go 12 blocks west." What is the straight-line distance between Gas Station Alpha and Junk Market Beta?
   (a) It is impossible to answer this without more information.
   (b) 1.3 km
   (c) 2.6 km
   (d) 3.4 km
   (e) 6.0 km

5. An American style pro football field is shaped like a rectangle, with sidelines that are perpendicular to the goal lines. Imagine that the two sidelines are extended indefinitely in both directions. The resulting pair of lines
   (a) would intersect at a distant point

(b) would intersect at two distant points

(c) would be skewed

(d) would be parallel

(e) would be perpendicular

6. An American style pro football field is shaped like a rectangle, with yard lines that are all parallel to the goal lines. Fill in the blank to make the following sentence true: "The yard lines on a football field are all _____ with respect to each other."

(a) perpendicular

(b) skewed

(c) equidistant

(d) parallel

(e) orthogonal

7. In the polar coordinate plane, the direction can be expressed in

(a) meters

(b) degrees

(c) square units

(d) cubic units

(e) linear units

8. Refer to Fig. Test 3-1. Suppose lines L, M, and N are all straight, and they all lie in the same plane. Suppose line M intersects line N at point P, and line L intersects line N at point Q. Further suppose that lines L and M are parallel. Let the angles of intersection, and their measures, be denoted as p_1 through p_4 and q_1 through q_4, as shown. Which of the following statements (a), (b), (c), or (d) is not necessarily true?

(a) $p_1 = q_1$

(b) $q_2 = q_3$

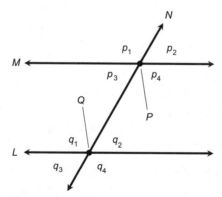

Fig. Test 3-1. Illustration for Part Three Test Questions 8 through 11.

(c) $q_1 = p_4$

(d) $q_1 = q_3$

(e) All of the statements (a) through (d) are true.

9. Refer to Fig. Test 3-1. Suppose lines L, M, and N are all straight, and they all lie in the same plane. Suppose line M intersects line N at point P, and line L intersects line N at point Q. Further suppose that lines L and M are parallel. Let the angles of intersection, and their measures, be denoted as p_1 through p_4 and q_1 through q_4, as shown. Angles p_3 and q_3 are called

(a) alternate interior angles

(b) alternate exterior angles

(c) vertical angles

(d) horizontal angles

(e) corresponding angles

10. Refer to Fig. Test 3-1. Suppose lines L, M, and N are all straight, and they all lie in the same plane. Suppose line M intersects line N at point P, and line L intersects line N at point Q. Further suppose that lines L and M are parallel. Let the angles of intersection, and their measures, be denoted as p_1 through p_4 and q_1 through q_4, as shown. Angles p_1 and p_4 are called

(a) alternate interior angles

(b) alternate exterior angles

(c) vertical angles

(d) horizontal angles

(e) corresponding angles

11. Refer to Fig. Test 3-1. Suppose lines L, M, and N are all straight, and they all lie in the same plane. Suppose line M intersects line N at point P, and line L intersects line N at point Q. Let the angles of intersection, and their measures, be denoted as p_1 through p_4 and q_1 through q_4, as shown. Suppose that $p_1 = q_1$. From this we can surmise that

(a) lines L and M are skew

(b) lines L and M are perpendicular

(c) lines L and M are parallel

(d) lines L and M intersect to the left of points P and Q

(e) lines L and M intersect to the right of points P and Q

12. Imagine a city laid out in square blocks on a level expanse of prairie, with "streets" running east–west and "avenues" running north–south. The distances between adjacent intersections are all

exactly 200 m. You stop at Restaurant Gamma and have supper, and on the way out you ask the cashier how to get to Motel Delta. The cashier says, "Go 8 blocks east, then turn right and go 6 blocks south." What is the distance you must drive to get to the motel?
(a) It is impossible to answer this without more information.
(b) 1.0 km
(c) 1.4 km
(d) 2.0 km
(e) 2.8 km

13. In Fig. Test 3-2, what is the distance between the labeled points $(-3, 4)$ and $(0, 0)$?
(a) 3 units
(b) 4 units
(c) 5 units
(d) 7 units
(e) It is impossible to determine this without more information.

14. In Fig. Test 3-2, what is the slope of the heavy, slanted line?
(a) $-4/3$
(b) $-3/4$
(c) $3/4$
(d) $4/3$
(e) It is impossible to determine this without more information.

15. In Fig. Test 3-2, what is the equation of the heavy, slanted line?
(a) $(-4/3)x + y = 0$
(b) $(-3/4)x + y = 0$

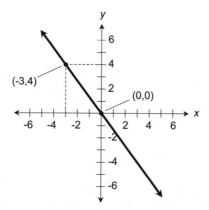

Fig. Test 3-2.　Illustration for Part Three Test Questions 13 through 16.

(c) $(3/4)x+y=0$

(d) $(4/3)x+y=0$

(e) It is impossible to determine this without more information.

16.　In Fig. Test 3-2, suppose L is a line perpendicular to the heavy, slanted line. What is the equation of L?

(a) $(-4/3)x+y=0$

(b) $(-3/4)x+y=0$

(c) $(3/4)x+y=0$

(d) $(4/3)x+y=0$

(e) It is impossible to determine this without more information.

17.　Suppose you live in a house that is shaped like a rectangular prism. It is exactly 15 m long, 20 m wide, and 5 m high, with a flat roof. There are two stories and no basement. Neglecting the volume taken up by the interior walls, floors, ceilings, furniture, and other objects, how much living room, in terms of volume, do you have?

(a) $75\,\text{m}^3$

(b) $100\,\text{m}^3$

(c) $300\,\text{m}^3$

(d) $600\,\text{m}^3$

(e) $1500\,\text{m}^3$

18.　Suppose you live in a house that is shaped like a rectangular prism. It is exactly 15 m long, 20 m wide, and 5 m high, with a flat roof. There are two stories and no basement. Neglecting the space taken up by the interior walls, furniture, and other objects, how much living room, in terms of floor area, do you have?

(a) $75\,\text{m}^2$

(b) $100\,\text{m}^2$

(c) $300\,\text{m}^2$

(d) $600\,\text{m}^2$

(e) $1500\,\text{m}^2$

19.　How many seconds of arc are there in a complete circle?

(a) 2π

(b) 360

(c) 21,600

(d) 1,296,000

(e) It is impossible to answer this without more information.

20.　Imagine a city laid out on a level expanse of prairie, with "streets" that are concentric circles spaced 200 m apart and "avenues" that go radially outward from the city center at angles that are all

multiples of 10° and that are numbered according to their angle counterclockwise from due east. You check out of Motel Delta and ask the clerk how to get to State Complex Epsilon. The attendant says, "Just follow this avenue outward 2 blocks, then turn left and travel counterclockwise exactly 9 blocks." How far must you drive?
(a) It is impossible to answer this question without more information.
(b) 0.9 km
(c) 1.1 km
(d) 1.8 km
(e) 2.2 km

21. A major league baseball infield is built to specific dimensions. Home plate, first base, second base, and third base are all at the vertices of a square that measures 90 feet (90 ft) on a side. The diagonal running from home plate to second base is the same length as the diagonal running from first base to third base. Let P be the point where these diagonals intersect. The pitcher's rubber lies along the diagonal from home plate to second base, and is 60.5 ft from home plate. Which of the following statements is true?
(a) The pitcher's rubber is closer to home plate than to second base.
(b) The pitcher's rubber is closer to second base than to home plate.
(c) The pitcher's rubber is equidistant from home plate and second base.
(d) The pitcher's rubber is centered at point P.
(e) None of the above.

22. Suppose you see the shadow of a post cast by the sun on a flat pavement. The cosine of the sun's angle above the horizon is equal to
(a) the height of the post, divided by the length of the post's shadow
(b) the length of the post's shadow, divided by the height of the post
(c) the height of the post, divided by the distance from the top of the post to the tip of its shadow
(d) the distance from the top of the post to the tip of its shadow, divided by the height of the post
(e) none of the above

23. According to Fig. Test 3-3, point A suggests that
(a) $\sin -90° = 1$
(b) $\sin 1 = -90°$
(c) $\sin -90 = -1°$

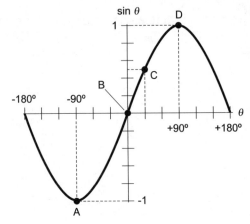

Fig. Test 3-3.   Illustration for Part Three Test Questions 23 through 28.

(d) $\sin -90° = -1$

(e) the equation of the graph is linear

24. Which of the following four statements (a), (b), (c), or (d), if any, does Fig. Test 3-3 fail to suggest?
    (a) $\sin -90° = -\sin +90°$
    (b) $\sin +90° = -\sin -90°$
    (c) $\sin 0° = 0$
    (d) $\sin +30° = 0.5$
    (e) The figure suggests that (a), (b), (c), and (d) are all true.

25. The extent of the domain of angles θ shown in Fig. Test 3-3, in radians, is
    (a) $-\pi/4 \le \theta \le +\pi/4$
    (b) $-\pi/2 \le \theta \le +\pi/2$
    (c) $-\pi \le \theta \le +\pi$
    (d) $-2\pi \le \theta \le +2\pi$
    (e) $-4\pi \le \theta \le +4\pi$

26. The graph of Fig. Test 3-3 suggests that the range of the sine function, when the independent variable is restricted to angles between and including $-180°$ and $+180°$, is
    (a) $-1 \le \sin \theta \le 1$
    (b) $0 \le \sin \theta \le 1$
    (c) $-1 \le \sin \theta \le 0$
    (d) $0 \le \sin \theta \le -1$
    (e) $1 \le \sin \theta \le -1$·

27. The curve shown in Fig. Test 3-3 is known as
    (a) a sinusoid
    (b) a circular wave
    (c) a polar wave
    (d) a radial wave
    (e) a Cartesian wave

28. The curve shown in Fig. Test 3-3 represents
    (a) 1/4 of a wave cycle
    (b) 1/2 of a wave cycle
    (c) one full wave cycle
    (d) two full wave cycles
    (e) four full wave cycles

29. Conventional time is measured according to the *mean solar day* *(MSD)*. For an observer in the northern hemisphere, the MSD is the average length of time it takes for the sun to "revolve once around" from a point directly in the south to a point directly in the south on the following day. This is, by convention, divided into 24 hours. This means that in 1 hour:
    (a) the earth rotates $1°$ on its axis with respect to the sun
    (b) the earth rotates $10°$ on its axis with respect to the sun
    (c) the earth rotates $15°$ on its axis with respect to the sun
    (d) the earth rotates $24°$ on its axis with respect to the sun
    (e) the earth rotates $30°$ on its axis with respect to the sun

30. Imagine a city laid out on a level expanse of prairie, with "streets" that are concentric circles spaced 200 m apart and "avenues" that go radially outward from the city center at angles that are all multiples of $10°$ and that are numbered according to their angle counterclockwise from due east. You have been telling your senator at State Complex Epsilon how much you hate her economic policy, and then you ask her how to get to Tera Mall Theta. She says, "Just follow this avenue outward 6 blocks." What is the straight-line distance between State Complex Epsilon and Tera Mall Theta?
    (a) It is impossible to answer this question without more information.
    (b) 0.6 km
    (c) 1.2 km
    (d) 1.8 km
    (e) 3.6 km

31. The Cartesian plane is an example of a
    (a) rectangular coordinate system
    (b) polar coordinate system
    (c) nonlinear coordinate system
    (d) circular coordinate system
    (e) spherical coordinate system

32. In Fig. Test 3-4, suppose that each of the faces is a parallelogram, and $s_1 = s_2 = s_3$. In this case, we can be certain that the figure is a
    (a) cube
    (b) skew prism
    (c) rectangular prism
    (d) hexagonal prism
    (e) parallelepiped

33. In Fig. Test 3-4, suppose angles x, y, and z each measure 90°. Also suppose that each of the faces is a parallelogram. In this case, we can be certain that the figure is a
    (a) cube
    (b) skew prism
    (c) rectangular prism
    (d) hexagonal prism
    (e) rhombus

34. In Fig. Test 3-4, suppose angles x, y, and z each measure 70°. Also suppose that each of the faces is a parallelogram. In this case, we can be certain that the figure is a
    (a) parallelepiped
    (b) rhombus
    (c) rectangular prism

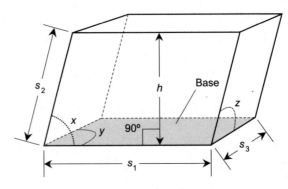

Fig. Test 3-4.   Illustration for Part Three Test Questions 32 through 35.

(d) hexagonal prism

(e) skew prism

35. In Fig. Test 3-4, suppose $h = 5\,\text{m}$ and the area of the base is $18\,\text{m}^2$. Also suppose that each of the faces is a parallelogram. What is the volume of the figure?

(a) $23\,\text{m}^3$

(b) $30\,\text{m}^3$

(c) $45\,\text{m}^3$

(d) $90\,\text{m}^3$

(e) It is impossible to determine this without more information.

36. Imagine that you have two spherical balls, called Ball A and Ball B. Suppose Ball B has four times the surface area of Ball A. What can we say about their relative diameters?

(a) The diameter of Ball B is twice the diameter of Ball A.

(b) The diameter of Ball B is four times the diameter of Ball A.

(c) The diameter of Ball B is eight times the diameter of Ball A.

(d) The diameter of Ball B is 16 times the diameter of Ball A.

(e) Nothing, without more information.

37. Suppose you are in a perfectly cubical room measuring exactly $3\,\text{m}$ high by $3\,\text{m}$ wide by $3\,\text{m}$ deep. You draw a straight line L along one wall (call it wall X) so that line L runs "kitty-corner" diagonally across the square formed by wall X, dividing wall X into two identical triangles. What is the length, in meters, of line L?

(a) The square root of 3.

(b) The square root of 6.

(c) The square root of 12.

(d) The square root of 18.

(e) The square root of 27.

38. Suppose you are in a perfectly cubical room measuring exactly $3\,\text{m}$ high by $3\,\text{m}$ wide by $3\,\text{m}$ deep. You draw a straight line L along one wall (call it wall X) so that line L runs "kitty-corner" diagonally across the square formed by wall X, dividing wall X into two identical triangles. What is the area of either one of these triangles?

(a) It is impossible to determine this without more information.

(b) $4.5\,\text{m}^2$

(c) $6\,\text{m}^2$

(d) $9\,\text{m}^2$

(e) $27\,\text{m}^2$

39.   Suppose you are in a perfectly cubical room measuring exactly 3 m high by 3 m wide by 3 m deep. You draw a straight line L along one wall (call it wall X) so that line L is parallel to the floor, and so that one point on line L is 2 m above the floor. From this, you can be sure that line L

(a) intersects the ceiling along wall X, but not necessarily at a right angle

(b) intersects the ceiling along some wall other than X, and not necessarily at a right angle

(c) intersects the ceiling along wall X at a right angle

(d) intersects the ceiling along some wall other than X at a right angle

(e) does not intersect the ceiling

40.   Suppose you are in a perfectly cubical room measuring exactly 3 m high by 3 m wide by 3 m deep. You draw a straight line L along one wall (call it wall X) so that line L intersects the floor at a right angle. From this, you can be sure that line L

(a) intersects the ceiling along wall X, but not necessarily at a right angle

(b) intersects the ceiling along some wall other than X, and not necessarily at a right angle

(c) intersects the ceiling along wall X at a right angle

(d) intersects the ceiling along some wall other than X at a right angle

(e) does not intersect the ceiling

PART FOUR

Math in Science

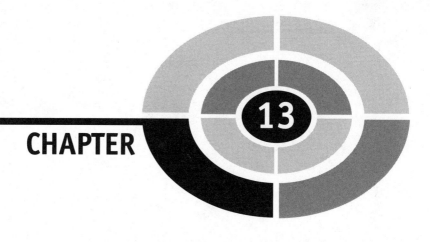

Vectors and 3D

Cartesian three-space, also called *rectangular three-space* or *xyz-space*, is defined by three number lines that intersect at a common *origin* point. At the origin, each of the three number lines is perpendicular to the other two. This makes it possible to pictorially relate one variable to another. Most three-dimensional (3D) graphs look line lines, curves, or surfaces. Renditions are enhanced by computer graphics programs.

You should know middle-school algebra to understand this chapter.

Vectors in the Cartesian Plane

A *vector* is a mathematical expression for a quantity with two independent properties: *magnitude* and *direction*. The direction, also called *orientation*, is defined in the sense of a ray, so it "points" somewhere. Vectors are used to represent physical variables such as distance, velocity, and acceleration. Conventionally, vectors are denoted by boldface letters of the alphabet.

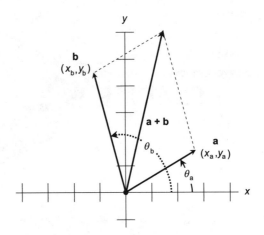

Fig. 13-1. Two vectors in the Cartesian plane. They are added using the "parallelogram method."

In the xy-plane, vectors **a** and **b** can be illustrated as rays from the origin (0,0) to points (x_a, y_a) and (x_b, y_b) as shown in Fig. 13-1.

EQUIVALENT VECTORS

Occasionally, a vector is expressed in a form that begins at a point other than the origin (0,0). In order for the following formulas to hold, such a vector must be reduced to *standard form*, so it "begins" at the origin. This can be accomplished by subtracting the coordinates (x_0, y_0) of the starting point from the coordinates of the end point (x_1, y_1). For example, if a vector **a*** starts at (3,−2) and ends at (1,−3), it reduces to an *equivalent vector* **a** in standard form:

$$a = \{(1 - 3), [-3 - (-2)]\}$$
$$= (-2, -1)$$

Any vector **a*** that is parallel to **a**, and that has the same length and the same direction (or orientation) as **a**, is equal to vector **a**. A vector is defined solely on the basis of its magnitude and its direction (or orientation). Neither of these two properties depends on the location of the end point.

MAGNITUDE

The magnitude (also called the *length, intensity,* or *absolute value*) of vector **a**, written |**a**| or a, can be found in the Cartesian plane by using a distance formula resembling the Pythagorean theorem:

$$|\mathbf{a}| = (x_a^2 + y_a^2)^{1/2}$$

Remember that the 1/2 power is the same thing as the square root.

DIRECTION

The direction of vector $\mathbf{a}$, written dir $\mathbf{a}$, is the angle θ_a at which vector $\mathbf{a}$ points, as measured counterclockwise from the positive x axis:

$$\text{dir } \mathbf{a} = \theta_a$$

From basic trigonometry, notice that $\tan \theta_a = y_a/x_a$. Therefore, θ_a is equal to the *inverse tangent*, also called the *arctangent* (abbreviated arctan or $\tan^{-1}$) of y_a/x_a. So:

$$\text{dir } \mathbf{a} = \theta_a = \arctan(y_a/x_a) = \tan^{-1}(y_a/x_a)$$

By convention, the angle θ_a is reduced to a value that is at least zero, but less than one full counterclockwise revolution. That is, $0° \leq \theta_a < 360°$ (if the angle is expressed in degrees), or $0 \text{ rad} \leq \theta_a < 2\pi \text{ rad}$ (if the angle is expressed in radians).

SUM

The sum of two vectors $\mathbf{a}$ and $\mathbf{b}$, where $\mathbf{a} = (x_a, y_a)$ and $\mathbf{b} = (x_b, y_b)$, can be found by adding their x values and y values independently:

$$\mathbf{a} + \mathbf{b} = [(x_a + x_b), (y_a + y_b)]$$

This sum can be found geometrically by constructing a parallelogram with $\mathbf{a}$ and $\mathbf{b}$ as adjacent sides. Then $\mathbf{a} + \mathbf{b}$ is the diagonal of this parallelogram (Fig. 13-1).

MULTIPLICATION BY SCALAR

To multiply a vector by a *scalar* (an ordinary real number), the x and y components of the vector are both multiplied by that scalar. If we have a vector $\mathbf{a} = (x_a, y_a)$ and a scalar k, then

$$k\mathbf{a} = \mathbf{a}k = (kx_a, ky_a)$$

When a vector is multiplied by a scalar, the length of the vector changes. If the scalar is positive, the direction of the product vector is the same as that of the original vector. If the scalar is negative, the direction of the product vector is opposite that of the original vector. If the scalar is 0, the result is known as the *zero vector*. It can be denoted by the ordered pair (0,0).

DOT PRODUCT

Let $\mathbf{a} = (x_a, y_a)$ and $\mathbf{b} = (x_b, y_b)$. The *dot product*, also known as the *scalar product* and written $\mathbf{a} \cdot \mathbf{b}$, of two vectors $\mathbf{a}$ and $\mathbf{b}$ is a number (that is, a scalar) that can be found using this formula:

$$\mathbf{a} \cdot \mathbf{b} = x_a x_b + y_a y_b$$

PROBLEM 13-1
What is the sum of $\mathbf{a} = (3, -5)$ and $\mathbf{b} = (2, 6)$?

SOLUTION 13-1
Add the x and y components together independently:

$$\mathbf{a} + \mathbf{b} = [(3 + 2), (-5 + 6)]$$
$$= (5, 1)$$

PROBLEM 13-2
What is the dot product of $\mathbf{a} = (3, -5)$ and $\mathbf{b} = (2, 6)$?

SOLUTION 13-2
Use the formula given above for the dot product:

$$\mathbf{a} \cdot \mathbf{b} = (3 \times 2) + (-5 \times 6)$$
$$= 6 + (-30)$$
$$= -24$$

PROBLEM 13-3
What happens if the order of the dot product is reversed? Does the value change?

SOLUTION 13-3
No. The value of the dot product of two vectors does not depend on the order in which the vectors are "dot-multiplied." This can be proven easily

from the formula above. Let $\mathbf{a} = (x_a, y_a)$ and $\mathbf{b} = (x_b, y_b)$. First consider the dot product of $\mathbf{a}$ and $\mathbf{b}$ (pronounced "$\mathbf{a}$ dot $\mathbf{b}$"):

$$\mathbf{a} \cdot \mathbf{b} = x_a x_b + y_a y_b$$

Now consider the dot product $\mathbf{b} \cdot \mathbf{a}$:

$$\mathbf{b} \cdot \mathbf{a} = x_b x_a + y_b y_a$$

Because ordinary multiplication is commutative – that is, the order in which the factors are multiplied doesn't matter – we can convert the above formula to this:

$$\mathbf{b} \cdot \mathbf{a} = x_a x_b + y_a y_b$$

But $x_a x_b + y_a y_b$ is just another way of writing $\mathbf{a} \cdot \mathbf{b}$. So, for any two vectors $\mathbf{a}$ and $\mathbf{b}$, it is always true that $\mathbf{a} \cdot \mathbf{b} = \mathbf{b} \cdot \mathbf{a}$.

Rectangular 3D Coordinates

Figure 13-2 illustrates the simplest possible set of *rectangular 3D coordinates*, also called *Cartesian three-space* or *xyz-space*. All three number lines

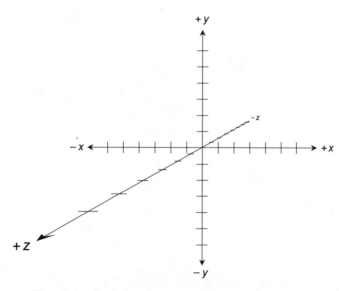

Fig. 13-2.   Cartesian three-space, also called *xyz*-space.

have equal increments. (This is a perspective illustration, so the increments on the z axis appear distorted. A true 3D rendition would have the positive z axis perpendicular to the page.) The three number lines all intersect at a single point, the origin, which corresponds to the zero points on each line.

The horizontal (right-and-left) axis is called the x axis; the vertical (up-and-down) axis is called the y axis, and the page-perpendicular (in-and-out) axis is called the z axis. In most renditions of rectangular 3D coordinates, the positive x axis runs from the origin toward the viewer's right, and the negative x axis runs toward the left. The positive y axis runs upward, and the negative y axis runs downward. The positive z axis "points out of the page directly at you," and the negative z axis "points behind the page straight away from you." Sometimes you will see graphs in which the y and z axes are interchanged from the way they're shown here, so the y axis is perpendicular to the page while the z axis runs up and down.

ORDERED TRIPLES AS POINTS

Figure 13-3 shows two specific points, called P and Q, plotted in Cartesian three-space. The coordinates of point P are $(-5,-4,3)$, and the coordinates

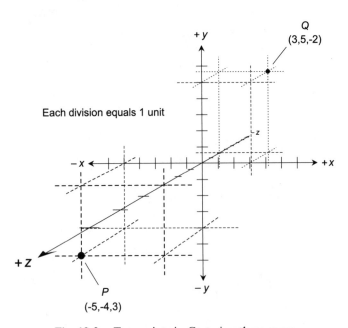

Fig. 13-3. Two points in Cartesian three-space.

of point Q are (3,5,−2). Points are denoted as *ordered triples* in the form (x,y,z), where the first number represents the value on the x axis, the second number represents the value on the y axis, and the third number represents the value on the z axis. The word "ordered" means that the order, or sequence, in which the numbers are listed is important. The ordered triple (1,2,3) is not the same as any of the ordered triples (1,3,2), (2,1,3), (2,3,1), (3,1,2), or (3,2,1), even though all of the triples contain the same three numbers.

In an ordered triple, there are no spaces after the commas, as there are in the notation of a set or sequence.

VARIABLES AND ORIGIN

In Cartesian three-space, there are usually two independent-variable coordinate axes and one dependent-variable axis. The x and y axes represent independent variables; the z axis represents a dependent variable whose value is affected by both the x and the y values.

In some scenarios, two of the variables are dependent and only one is independent. Most often, the independent variable in such cases is x. Rarely, you'll come across a situation in which none of the values depends on either of the other two, or when a correlation, but not a true relation, exists among the values of two or all three of the variables. Plots of this sort usually look like "swarms of points," representing the results of observations, or values predicted by some scientific theory. These graphs are known as *scatter plots*, and are common in statistics.

DISTANCE BETWEEN POINTS

Suppose there are two different points $P=(x_0,y_0,z_0)$ and $Q=(x_1,y_1,z_1)$ in Cartesian three-space. The distance d between these two points can be found using this formula:

$$d = [(x_1 - x_0)^2 + (y_1 - y_0)^2 + (z_1 - z_0)^2]^{1/2}$$

PROBLEM 13-4
What is the distance between the points $P=(-5,-4,3)$ and $Q=(3,5,-2)$ illustrated in Fig. 13-3? Express the answer rounded off to three decimal places.

SOLUTION 13-4

We can plug the coordinate values into the distance equation, where:

$$x_0 = -5$$
$$x_1 = 3$$
$$y_0 = -4$$
$$y_1 = 5$$
$$z_0 = 3$$
$$z_1 = -2$$

Therefore:

$$
\begin{aligned}
d &= \{[3-(-5)]^2 + [5-(-4)]^2 + (-2-3)^2\}^{1/2} \\
&= [8^2 + 9^2 + (-5)^2]^{1/2} \\
&= (64 + 81 + 25)^{1/2} \\
&= 170^{1/2} \\
&= 13.038
\end{aligned}
$$

Vectors in Cartesian Three-Space

A *vector* in Cartesian three-space is the same as a vector in the Cartesian plane, except that there is more "freedom" in terms of direction. This makes the expression of direction in 3D more complicated than is the case in 2D. It also makes vector arithmetic a lot more interesting!

EQUIVALENT VECTORS

In Cartesian three-space, vectors **a** and **b** can be denoted as arrow-tipped line segments from the origin $(0,0,0)$ to points (x_a, y_a, z_a) and (x_b, y_b, z_b), as shown in Fig. 13-4. This, like all three-space drawings in this chapter, is a perspective illustration. Both vectors in this example point in directions on the reader's side of the plane containing the page. In a true 3D model, both of them would "stick up out of the paper at an angle."

In Fig. 13-4, both vectors **a** and **b** have their end points at the origin. This is the standard form of a vector in any coordinate system. In order for the following formulas to hold, vectors must be expressed in standard form.

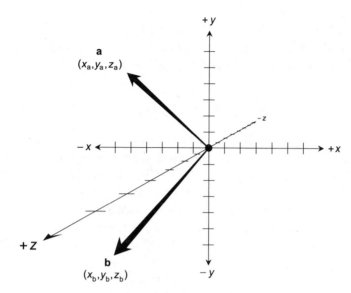

Fig. 13-4.   Vectors in *xyz*-space. This is a perspective drawing; both vectors point in
directions on your side of the plane containing the page.

If a given vector is not in standard form, it can be converted by subtracting
the coordinates (x_0,y_0,z_0) of the starting point from the coordinates of the
end point (x_1,y_1,z_1). For example, if a vector **a*** starts at (4,7,0) and ends
at (1,–3,5), it reduces to an equivalent vector **a** in standard form:

$$\mathbf{a} = [(1-4),(-3-7),(5-0)]$$
$$= (-3,-10,5)$$

Any vector **a***, which is parallel to **a** and has the same length as **a**, is equal
to vector **a**, because **a*** has the same magnitude and the same direction as **a**.
Similarly, any vector **b***, which is parallel to **b** and has the same length as **b**, is
defined as being equal to **b**. As in the 2D case, a vector is defined solely on
the basis of its magnitude and its direction. Neither of these two properties
depends on the location of the end point.

DEFINING THE MAGNITUDE

When the end point of a vector **a** is at the origin, the magnitude of **a**, written
|**a**| or *a*, can be found by a three-dimensional extension of the Pythagorean
theorem for right triangles. The formula looks like this:

$$|\mathbf{a}| = (x_a^2 + y_a^2 + z_a^2)^{1/2}$$

The magnitude of any vector **a** in standard form is equal to the distance of the end point from the origin. Note that the above formula is the distance formula for two points, (0,0,0) and (x_a, y_a, z_a).

DIRECTION ANGLES AND COSINES

The direction of a vector **a** in standard form can be defined by specifying the angles θ_x, θ_y, and θ_z that the vector **a** subtends relative to the positive x, y, and z axes respectively (Fig. 13-5). These angles, expressed in radians as an ordered triple $(\theta_x, \theta_y, \theta_z)$, are the *direction angles* of **a**.

Sometimes the cosines of these angles are used to define the direction of a vector **a** in 3D space. These are the *direction cosines* of **a**:

$$\text{dir } \mathbf{a} = (\alpha, \beta, \gamma)$$
$$\alpha = \cos\theta_x$$
$$\beta = \cos\theta_y$$
$$\gamma = \cos\theta_z$$

For any vector **a** in Cartesian three-space, the sum of the squares of the direction cosines is always equal to 1. That is,

$$\alpha^2 + \beta^2 + \gamma^2 = 1$$

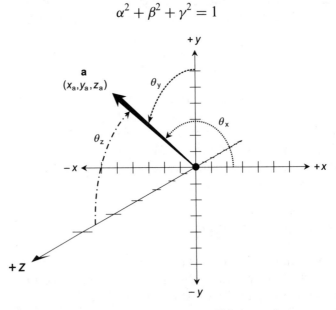

Fig. 13-5.   Direction angles of a vector in *xyz*-space. This is another perspective drawing; the vector points in a direction on your side of the plane containing the page.

Another way of expressing this is:

$$\cos^2 \theta_x + \cos^2 \theta_y + \cos^2 \theta_z = 1$$

where the expression $\cos^2 \theta$ means $(\cos \theta)^2$.

SUM

The sum of vectors $\mathbf{a} = (x_a, y_a, z_a)$ and $\mathbf{b} = (x_b, y_b, z_b)$ in standard form is given by the following formula:

$$\mathbf{a} + \mathbf{b} = [(x_a + x_b), (y_a + y_b), (z_a + z_b)]$$

This sum can be found geometrically by constructing a parallelogram in 3D space with $\mathbf{a}$ and $\mathbf{b}$ as adjacent sides. The sum $\mathbf{a} + \mathbf{b}$ is the diagonal of the parallelogram. This is shown in Fig. 13-6. (The parallelogram appears distorted because of the perspective of the drawing.)

MULTIPLICATION BY SCALAR

In three-dimensional Cartesian coordinates, let vector $\mathbf{a}$ be defined by the coordinates (x_a, y_a, z_a) when reduced to standard form. Suppose $\mathbf{a}$ is

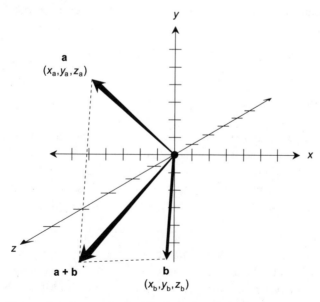

Fig. 13-6.   Vectors in *xyz*-space are added using the "parallelogram method." This is a perspective drawing, so the parallelogram appears distorted.

multiplied by a positive real-number scalar k. Then the following equation holds:

$$k\mathbf{a} = k(x_a, y_a, z_a) = (kx_a, ky_a, kz_a)$$

If $\mathbf{a}$ is multiplied by a negative real scalar $-k$, then:

$$-k\mathbf{a} = -k(x_a, y_a, z_a) = (-kx_a, -ky_a, -kz_a)$$

Suppose the direction angles of $\mathbf{a}$ are represented by the ordered triple $(\theta_{xa}, \theta_{ya}, \theta_{za})$. Then the direction angles of $k\mathbf{a}$ are the same; they are also $(\theta_{xa}, \theta_{ya}, \theta_{za})$. The direction angles of $-k\mathbf{a}$ are all changed by 180° (π rad). The direction angles of $-k\mathbf{a}$ can be found by adding or subtracting 180° (π rad) to or from each of the direction angles for $k\mathbf{a}$, so that the resulting angles all have measures of at least 0° (0 rad) but less than 360° (2π rad).

DOT PRODUCT

The *dot product*, also known as the *scalar product* and written $\mathbf{a} \cdot \mathbf{b}$, of two 3D vectors $\mathbf{a} = (x_a, y_a, z_a)$ and $\mathbf{b} = (x_b, y_b, z_b)$ in standard form is a real number given by the formula:

$$\mathbf{a} \cdot \mathbf{b} = x_a x_b + y_a y_b + z_a z_b$$

The dot product can also be found from the magnitudes $|\mathbf{a}|$ and $|\mathbf{b}|$, and the angle θ between vectors $\mathbf{a}$ and $\mathbf{b}$ as measured counterclockwise in the plane containing them both:

$$\mathbf{a} \cdot \mathbf{b} = |\mathbf{a}||\mathbf{b}| \cos \theta$$

CROSS PRODUCT

The *cross product*, also known as the *vector product* and written $\mathbf{a} \times \mathbf{b}$, of vectors $\mathbf{a} = (x_a, y_a, z_a)$ and $\mathbf{b} = (x_b, y_b, z_b)$ in standard form is a vector perpendicular to the plane containing both $\mathbf{a}$ and $\mathbf{b}$. Let θ be the angle between vectors $\mathbf{a}$ and $\mathbf{b}$ as measured counterclockwise in the plane containing them both, as shown in Fig. 13-7. The magnitude of $\mathbf{a} \times \mathbf{b}$ is given by the formula:

$$|\mathbf{a} \times \mathbf{b}| = |\mathbf{a}||\mathbf{b}| \sin \theta$$

In the example shown, $\mathbf{a} \times \mathbf{b}$ points upward at a right angle to the plane containing both vectors $\mathbf{a}$ and $\mathbf{b}$. If $0° < \theta < 180°$ (0 rad $< \theta < \pi$ rad), you can use the *right-hand rule* to ascertain the direction of $\mathbf{a} \times \mathbf{b}$. Curl your fingers in the direction that θ, the angle from $\mathbf{a}$ to $\mathbf{b}$, is defined. Extend your thumb. Then $\mathbf{a} \times \mathbf{b}$ points in the direction of your thumb.

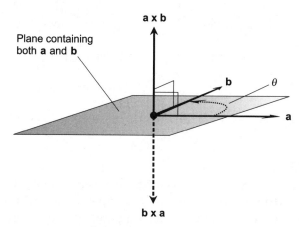

Fig. 13-7. The vector **b** × **a** has the same magnitude as vector **a** × **b**, but points in the opposite direction. Both cross products are perpendicular to the plane containing the two original vectors.

When $180° < \theta < 360°$ (π rad $< \theta < 2\pi$ rad), the cross-product vector reverses direction. Theoretically, the cross-product vector's direction angles do not change; but its magnitude becomes negative. This is demonstrated by the fact that, in the above formula, $\sin \theta$ is positive when $0° < \theta < 180°$ (0 rad $< \theta < \pi$ rad), but negative when $180° < \theta < 360°$ (π rad $< \theta < 2\pi$ rad).

UNIT VECTORS

Any vector **a**, reduced to standard form so its starting point is at the origin, ends up at some point (x_a, y_a, z_a). This vector can be broken down into the sum of three mutually perpendicular vectors, each of which lies along one of the coordinate axes as shown in Fig. 13-8:

$$\mathbf{a} = (x_a, y_a, z_a)$$
$$= (x_a, 0, 0) + (0, y_a, 0) + (0, 0, z_a)$$
$$= x_a(1, 0, 0) + y_a(0, 1, 0) + z_a(0, 0, 1)$$

The vectors $(1,0,0)$, $(0,1,0)$, and $(0,0,1)$ are called *unit vectors* because their length is 1. It is customary to name these vectors **i**, **j**, and **k**, as follows:

$$(1,0,0) = \mathbf{i}$$
$$(0,1,0) = \mathbf{j}$$
$$(0,0,1) = \mathbf{k}$$

Therefore, the vector **a** shown in Fig. 13-8 breaks down this way:

$$\mathbf{a} = (x_a, y_a, z_a) = x_a\mathbf{i} + y_a\mathbf{j} + z_a\mathbf{k}$$

PROBLEM 13-5

Break the vector $\mathbf{b} = (-2, 3, -7)$ down into a sum of multiples of the unit vectors **i**, **j**, and **k**.

SOLUTION 13-5

This is a simple process, but envisioning it requires a keen "mind's eye." If you have any trouble seeing this in your imagination, think of **i** as "one unit of width going to the right," **j** as "one unit of height going up," and **k** as "one unit of depth coming towards you." Here we go:

$$\mathbf{b} = (-2, 3, -7)$$

$$= -2 \times (1,0,0) + 3 \times (0,1,0) + [-7 \times (0,0,1)]$$

$$= -2\mathbf{i} + 3\mathbf{j} + (-7)\mathbf{k}$$

$$= -2\mathbf{i} + 3\mathbf{j} - 7\mathbf{k}$$

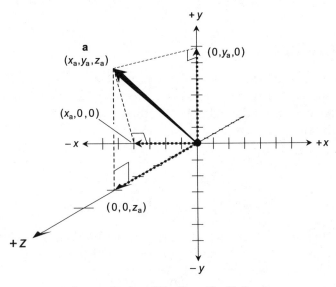

$$(x_a, y_a, z_a) = (x_a, 0, 0) + (0, y_a, 0) + (0, 0, z_a)$$

Fig. 13-8.   Any vector in Cartesian three-space can be broken up into a sum of three component vectors, each of which lies on one of the coordinate axes.

Flat Planes in Space

The equation of a flat geometric plane in Cartesian 3D coordinates is somewhat like the equation of a straight line in Cartesian 2D coordinates.

CRITERIA FOR UNIQUENESS

A flat geometric plane in 3D space can be uniquely defined according to any of the following criteria:

- Three points that do not all lie on the same straight line.
- A point in the plane and a vector normal (perpendicular) to the plane.
- Two intersecting straight lines.
- Two parallel straight lines.

GENERAL EQUATION OF PLANE

The simplest equation for a plane is derived on the basis of the second of the foregoing criteria: a point in the plane and a vector normal to the plane. Figure 13-9 shows a plane W in Cartesian three-space, a point

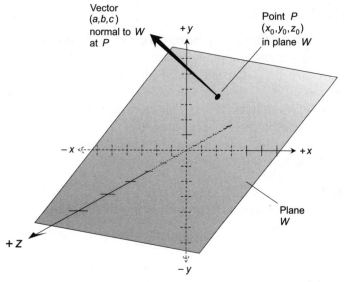

Fig. 13-9.   A plane W can be uniquely defined on the basis of a point P in the plane and a vector (a,b,c) normal to the plane. Dashed portions of the coordinate axes are "behind" the plane.

$P = (x_0, y_0, z_0)$ in plane W, and a *normal vector* $(a,b,c) = a\mathbf{i} + b\mathbf{j} + c\mathbf{k}$ that is perpendicular to plane W. The normal vector (a,b,c) in this illustration is shown originating at point P (rather than at the origin), because this particular plane W doesn't pass through the origin $(0,0,0)$. The values $x = a, y = b$, and $z = c$ for the vector are nevertheless based on its standard form.

When these things about a plane are known, we have enough information to uniquely define it and write its equation as follows:

$$a(x - x_0) + b(y - y_0) + c(z - z_0) = 0$$

In this form of the equation for a plane, the constants a, b, and c are called the *coefficients*. The above equation can also be written in this form:

$$ax + by + cz + d = 0$$

Here, the value of d is:

$$d = -(ax_0 + by_0 + cz_0)$$
$$= -ax_0 - by_0 - cz_0$$

PLOTTING A PLANE

In order to draw a graph of a plane based on its equation, it is good enough to know the points where the plane crosses each of the three coordinate axes. The plane can then be visualized, based on these points.

Not all planes cross all three of the axes in Cartesian xyz-space. If a plane is parallel to one of the axes, it does not cross that axis. If a plane is parallel to the plane formed by two of the three axes, then it crosses only the axis to which it is not parallel. But any plane in Cartesian 3D space must cross at least one of the coordinate axes at some point.

PROBLEM 13-6
Draw a graph of the plane W represented by the following equation:

$$-2x - 4y + 3z - 12 = 0$$

SOLUTION 13-6
The x-intercept, or the point where the plane W intersects the x axis, can be found by setting $y = 0$ and $z = 0$, and then solving for x. Call this

point P:

$$-2x - (4 \times 0) + (3 \times 0) - 12 = 0$$
$$-2x - 12 = 0$$
$$-2x = 12$$
$$x = 12/(-2) = -6$$
$$\text{Therefore } P = (-6,0,0)$$

The y-intercept, or the point where the plane W intersects the y axis, can be found by setting $x=0$ and $z=0$, and then solving for y. Call this point Q:

$$(-2 \times 0) - 4y + (3 \times 0) - 12 = 0$$
$$-4y - 12 = 0$$
$$-4y = 12$$
$$y = 12/(-4) = -3$$
$$\text{Therefore } Q = (0,-3,0)$$

The z-intercept, or the point where the plane W intersects the z axis, can be found by setting $x=0$ and $y=0$, and then solving for z. Call this point R:

$$(-2 \times 0) - (4 \times 0) + 3z - 12 = 0$$
$$3z - 12 = 0$$
$$3z = 12$$
$$z = 12/3 = 4$$
$$\text{Therefore } R = (0,0,4)$$

These three points are shown in the plot of Fig. 13-10. The plane can be envisioned, based on this data.

Note that some parts of the coordinate axes in Fig. 13-10 appear as dashed lines. This shows that these parts of the axes are "behind" the plane, according to the point of view from which we see the situation.

PROBLEM 13-7
Suppose a plane contains the point $(2,-7,0)$, and a normal vector to the plane at this point is $3\mathbf{i} + 3\mathbf{j} + 2\mathbf{k}$. What is the equation of this plane?

SOLUTION 13-7
The vector $3\mathbf{i} + 3\mathbf{j} + 2\mathbf{k}$ is equivalent to $(a,b,c) = (3,3,2)$. We have one point $(x_0,y_0,z_0) = (2,-7,0)$. Plugging these values into the general formula for

Each division equals 1 unit

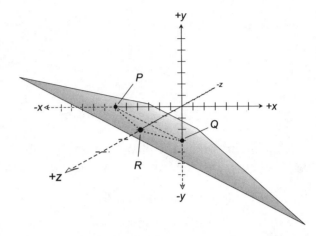

Fig. 13-10. Illustration for Problem 13-6. Dashed portions of the coordinate axes are "behind" the plane.

the equation of a plane gives us the following:

$$a(x - x_0) + b(y - y_0) + c(z - z_0) = 0$$

$$3(x - 2) + 3[y - (-7)] + 2(z - 0) = 0$$

$$3(x - 2) + 3(y + 7) + 2z = 0$$

$$3x - 6 + 3y + 21 + 2z = 0$$

$$3x + 3y + 2z + 15 = 0$$

Straight Lines in Space

Straight lines in Cartesian three-space present a more complicated picture than straight lines in the Cartesian coordinate plane. This is because there is an added dimension, making the expression of the direction more complex. But all linear equations, no matter what the number of dimensions, have one thing in common: they can be reduced to a form where no variable is raised to any power other than 0 or 1.

SYMMETRIC-FORM EQUATION

A straight line in Cartesian three-space can be represented by a "three-way" equation in three variables. This equation is known as a *symmetric-form equation*. It takes the following form, where x, y, and z are the variables, (x_0, y_0, z_0) represents the coordinates of a specific point on the line, and a, b, and c are constants:

$$(x - x_0)/a = (y - y_0)/b = (z - z_0)/c$$

It's important that none of the three constants a, b, or c be equal to zero. If $a = 0$ or $b = 0$ or $c = 0$, the result is a zero denominator in one of the expressions, and division by zero is not defined.

DIRECTION NUMBERS

In the symmetric-form equation of a straight line, the constants a, b, and c are known as the *direction numbers*. If we consider a vector $\mathbf{m}$ with its end point at the origin and its "arrowed end" at the point $(x,y,z) = (a,b,c)$, then the vector $\mathbf{m}$ is parallel to the line denoted by the symmetric-form equation. We have:

$$\mathbf{m} = a\mathbf{i} + b\mathbf{j} + c\mathbf{k}$$

where $\mathbf{m}$ is the three-dimensional equivalent of the slope of a line in the Cartesian plane. This is shown in Fig. 13-11 for a line L containing a point $P = (x_0, y_0, z_0)$.

PARAMETRIC EQUATIONS

There are infinitely many vectors that can satisfy the requirement for $\mathbf{m}$. (Just imagine $\mathbf{m}$ being longer or shorter, but still pointing in the same direction as it does in Fig. 13-11, or else in the exact opposite direction. All such vectors are parallel to line L.) If t is any nonzero real number, then $t\mathbf{m} = (ta, tb, tc) = ta\mathbf{i} + tb\mathbf{j} + tc\mathbf{k}$ will work just as well as $\mathbf{m}$ for the purpose of defining the direction of a line L. This gives us an alternative form for the equation of a line in Cartesian three-space:

$$x = x_0 + at$$
$$y = y_0 + bt$$
$$z = z_0 + ct$$

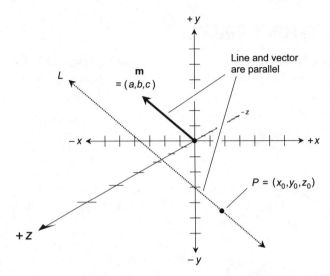

Fig. 13-11. A line L can be uniquely defined on the basis of a point P on the line and a vector $\mathbf{m} = (a,b,c)$ that is parallel to the line.

The nonzero real number t is called a *parameter*, and the above set of equations is known as a set of *parametric equations* for a straight line in *xyz*-space. In order for a complete geometric line (straight, and infinitely long) to be defined on this basis of parametric equations, the parameter t must be allowed to range over the entire set of real numbers, including zero.

PROBLEM 13-8

Find the symmetric-form equation for the line L shown in Fig. 13-12.

SOLUTION 13-8

The line L passes through the point $P = (-5,-4,3)$ and is parallel to the vector $\mathbf{m} = 3\mathbf{i} + 5\mathbf{j} - 2\mathbf{k}$. The direction numbers of L are the coefficients of the vector $\mathbf{m}$, that is:

$$a = 3$$
$$b = 5$$
$$c = -2$$

We are given a point P on L such that:

$$x_0 = -5$$
$$y_0 = -4$$
$$z_0 = 3$$

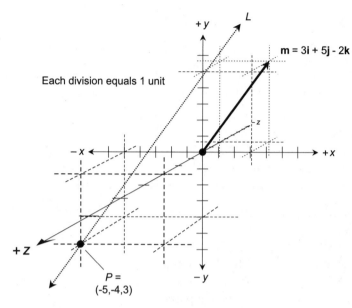

Fig. 13-12.   Illustration for Problems 13-8 and 13-9.

Plugging these values into the general symmetric-form equation for a line in Cartesian three-space gives us this:

$$(x - x_0)/a = (y - y_0)/b = (z - z_0)/c$$

$$[x - (-5)]/3 = [y - (-4)]/5 = (z - 3)/(-2)$$

$$(x + 5)/3 = (y + 4)/5 = (z - 3)/(-2)$$

PROBLEM 13-9

Find a set of parametric equations for the line L shown in Fig. 13-12.

SOLUTION 13-9

This involves nothing more than rearranging the values of $x_0, y_0, z_0, a, b,$ and c in the symmetric-form equation, and rewriting the data in the form of parametric equations. The results are:

$$x = -5 + 3t$$

$$y = -4 + 5t$$

$$z = 3 - 2t$$

Quiz

Refer to the text in this chapter if necessary. A good score is eight correct. Answers are in the back of the book.

1. The dot product $(3,5,0) \cdot (-4,-6,2)$ is equal to
    (a) the scalar quantity -4
    (b) the vector $(-12,-30,0)$
    (c) the scalar quantity -42
    (d) a vector perpendicular to the plane containing them both

2. What does the graph of the equation $y=3$ look like in Cartesian three-space?
    (a) A plane perpendicular to the y axis.
    (b) A plane parallel to the xy plane.
    (c) A line parallel to the y axis.
    (d) A line parallel to the xy plane.

3. Suppose vector **d** in the Cartesian plane begins at exactly $(1,1)$ and ends at exactly $(4,0)$. What is dir **d**, expressed to the nearest degree?
    (a) $342°$
    (b) $18°$
    (c) $0°$
    (d) $90°$

4. Suppose a line is represented by the equation $(x-3)/2 = (y+4)/5 = z-1$. Which of the following is a point on this line?
    (a) $(-3,4,-1)$
    (b) $(3,-4,1)$
    (c) $(2,5,1)$
    (d) There is no way to determine such a point without more information.

5. Suppose a vector $\mathbf{m} = a\mathbf{i} + b\mathbf{j} + c\mathbf{k}$ is parallel to a straight line L. What do the constants a, b, and c constitute?
    (a) Direction angles.
    (b) Direction numbers.
    (c) Direction cosines.
    (d) Direction vectors.

6. What can be said about the vectors $\mathbf{a}=(1,1,1)$ and $\mathbf{b}=(-2,-2,-2)$?
    (a) Vector **b** is a real-number multiple of vector **a**.
    (b) Vector **b** is oriented precisely opposite from vector **a**.

(c) Vectors **a** and **b** lie along the same straight line.

(d) All of the above statements are true.

7. What is the cross product $(2\mathbf{i} + 0\mathbf{j} + 0\mathbf{k}) \times (0\mathbf{i} + 2\mathbf{j} + 0\mathbf{k})$?
    (a) $0\mathbf{i} + 0\mathbf{j} + 0\mathbf{k}$
    (b) $2\mathbf{i} + 2\mathbf{j} + 0\mathbf{k}$
    (c) $0\mathbf{i} + 0\mathbf{j} + 4\mathbf{k}$
    (d) The scalar 0.

8. What is the sum of the two vectors $(3,5)$ and $(-5,-3)$?
    (a) $(0,0)$
    (b) $(8,8)$
    (c) $(2,2)$
    (d) $(-2,2)$

9. If a straight line in Cartesian three-space has direction defined by $\mathbf{m} = 0\mathbf{i} + 0\mathbf{j} + 3\mathbf{k}$, we can surmise:
    (a) that the line is parallel to the x axis
    (b) that the line lies in the yz plane
    (c) that the line lies in the xy plane
    (d) none of the above

10. Suppose a plane passes through the origin, and a vector normal to the plane is represented by $4\mathbf{i} - 5\mathbf{j} + 8\mathbf{k}$. The equation of this plane is
    (a) $4x - 5y + 8z = 0$
    (b) $-4x + 5y - 8z = 7$
    (c) $(x - 4) = (y + 5) = (z - 8)$
    (d) impossible to determine without more information

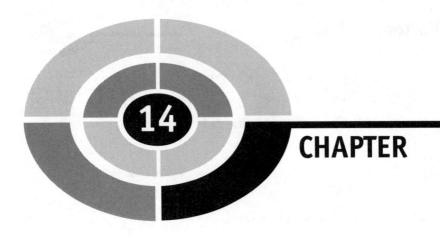

CHAPTER

14

Growth and Decay

Rates of change can sometimes be expressed in terms of mathematical constants. In this chapter, we'll examine a few of the ways this can happen.

Growth by Addition

The simplest changeable quantities can be written as lists of numbers whose values repeatedly increase or decrease by a fixed amount. Here are some examples:

$$A = 1, 2, 3, 4, 5, 6$$
$$B = 0, -1, -2, -3, -4, -5$$
$$C = 2, 4, 6, 8$$
$$D = -5, -10, -15, -20$$
$$E = 4, 8, 12, 16, 20, 24, 28, \ldots$$
$$F = 2, 0, -2, -4, -6, -8, -10, \ldots$$

The first four of these sequences are finite. The last two are infinite, as indicated by the three dots following the last term in each case.

ARITHMETIC PROGRESSION

In each of the six sequences shown above, the values either increase (A, C, and E) or else they decrease (B, D, and F). In all six sequences, the "spacing" between numbers is constant throughout. Note:

- The values in A always increase by 1.
- The values in B always decrease by 1.
- The values in C always increase by 2.
- The values in D always decrease by 5.
- The values in E always increase by 4.
- The values in F always decrease by 2.

Each sequence has a starting point or first number. After that, succeeding numbers can be predicted by repeatedly adding a constant. If the added constant is positive, the sequence increases. If the added constant is negative, the sequence decreases.

Let s_0 be the first number in a sequence S, and let c be a constant. Imagine that S can be written in this form:

$$S = s_0, (s_0 + c), (s_0 + 2c), (s_0 + 3c), \ldots$$

for as far as the sequence happens to go. Such a sequence is called an *arithmetic sequence* or an *arithmetic progression*. In this context, the word "arithmetic" is pronounced "air-ith-MET-ick."

The numbers s_0 and c can be whole numbers, but that is not a requirement. They can be fractions such as 2/3 or −7/5. They can be irrational numbers such as the square root of 2. As long as the separation between any two adjacent terms in a sequence is the same, the sequence is an arithmetic progression. In fact, even if s_0 and c are both equal to 0, the resulting sequence is an arithmetic progression.

ARITHMETIC SERIES

A *series* is, by definition, the sum of all the terms in a sequence. For any arithmetic sequence, the corresponding *arithmetic series* can be defined only if the sequence is finite. That means it must have a finite number of terms. For the above sequences A through F, let the corresponding series be called A_+

through F_+. Then:

$$A_+ = 1 + 2 + 3 + 4 + 5 + 6 = 21$$
$$B_+ = 0 + (-1) + (-2) + (-3) + (-4) + (-5) = -15$$
$$C_+ = 2 + 4 + 6 + 8 = 20$$
$$D_+ = (-5) + (-10) + (-15) + (-20) = -50$$

E_+ is not defined

F_+ is not defined

ARITHMETIC INTERPOLATION

When you see a long sequence of numbers, you should be able to tell without much trouble whether or not it's an arithmetic sequence. If it isn't immediately obvious, you can conduct a test: subtract each number from the one after it. If all the differences are the same, then the sequence is an arithmetic sequence.

Imagine that you see a long sequence of numbers, and some of the intermediate values are missing. An example of such a situation is shown in Table 14-1. It's not too hard to figure out what the missing values are, once you realize that this is an arithmetic sequence in which each term has a value that is 3 larger than the term preceding it. The 4th term has a value of 14, and the 9th term has a value of 29. The process of filling in missing intermediate values in a sequence is a form of *interpolation*. We might call the process of filling in Table 14-1 *arithmetic interpolation*.

ARITHMETIC EXTRAPOLATION

It is possible to "predict" what the next numbers are, or would be if a series were lengthened. Table 14-2 illustrates an example of this type of situation.

Table 14-1   A sequence with some intermediate values missing. They can be filled in by interpolation.

Position:	1st	2nd	3rd	4th	5th	6th	7th	8th	9th	10th	11th
Value:	5	8	11	?	17	20	23	26	?	32	35

Table 14-2 Extension of an arithmetic sequence can be done by extrapolation.

Position:	1st	2nd	3rd	4th	5th	6th	7th	8th	9th	10th	11th
Value:	32	30	28	26	24	22	20	?	?	?	?

The values keep getting smaller by 2. The 7th number is 20. Therefore, the 8th number must be 18, the 9th number must be 16, the 10th number must be 14, and the 11th number must be 12. This extension process constitutes a form of *extrapolation*, so we can call the process of filling in Table 14-2 *arithmetic extrapolation*.

PLOTTING AN ARITHMETIC SEQUENCE

An arithmetic sequence looks like a set of points when plotted in Cartesian (rectangular) coordinates. The term number or position in the sequence is depicted on the horizontal axis, and it plays the role of the independent variable. The term value is plotted against the vertical axis, and it plays the role of the dependent variable.

Figure 14-1 shows examples of two arithmetic sequences as they appear when graphed. (The dashed lines connect the dots, but they aren't actually parts of the sequences.) Note that the dashed lines are straight. One sequence is increasing, and the dashed line connecting this set of points ramps upward as you go toward the right. Because this sequence is finite, the dashed line ends at the final point (6,6). The other sequence is decreasing, and the dashed line for it ramps downward as you go toward the right. This sequence is infinite, as shown by the three dots at the end of the string of numbers, and also by the arrow at the right-hand end of the dashed line.

When any arithmetic sequence is graphed in this way, its points lie along a straight line. The slope m of the line depends on whether the sequence increases (positive slope) or decreases (negative slope). In fact, m is exactly equal to the constant c.

BE CAREFUL!

When interpolating or extrapolating any sequence, you must be sure to check out enough values to know that you're looking at a true sequence, and not just a string of numbers. If the string has only two or three values that increase by the same amount, you don't have enough information to be

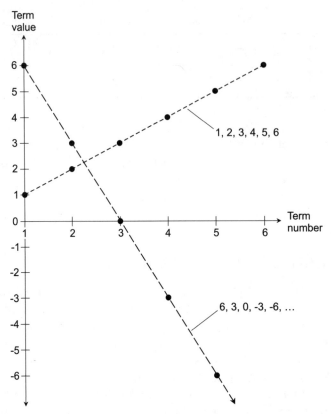

Fig. 14-1. When the values of the terms in an arithmetic sequence are plotted in rectangular coordinates, the points always fall along straight lines.

sure you're observing an arithmetic sequence. You need at least four or five values.

There's another way to go wrong when attempting to interpolate a sequence. If the missing values are alternate numbers, you can be deceived. The examples in Tables 14-3A, B, and C provide a vivid illustration of how far astray you can go if you jump to conclusions about filling in a sequence with alternate terms missing! The plot in Fig. 14-2 illustrates this graphically.

PROBLEM 14-1

Consider again the general formula for an arithmetic sequence S:

$$S = s_0, (s_0 + c), (s_0 + 2c), (s_0 + 3c), \ldots$$

Suppose $s_0 = 5$ and $c = 3$. List the first 10 terms of an infinite sequence that follows this form and has these values.

Table 14-3

(A) Suppose you see a sequence with alternate values missing, like this:

Position:	1st	2nd	3rd	4th	5th	6th	7th	8th	9th	10th	11th
Value:	1		2		3		4		5		6

(B) It's tempting to think that it is an arithmetic sequence, and that the missing values should be filled in to get this:

Position:	1st	2nd	3rd	4th	5th	6th	7th	8th	9th	10th	11th
Value:	1	1.5	2	2.5	3	3.5	4	4.5	5	5.5	6

(C) But in reality, it is not an arithmetic sequence. Instead, it goes like this:

Position:	1st	2nd	3rd	4th	5th	6th	7th	8th	9th	10th	11th
Value:	1	−1	2	−2	3	−3	4	−4	5	−5	6

SOLUTION 14-1

The first number is 5, and the numbers increase by 3 every time thereafter. So:

$$S = 5, 8, 11, 14, 17, 20, 23, 26, 29, 32, \ldots$$

PROBLEM 14-2

Is the following sequence an arithmetic sequence? If so, what are the values s_0 (the starting value) and c (the constant of change)?

$$S = 2, 4, 8, 16, 32, 64, \ldots$$

SOLUTION 14-2

This is not an arithmetic sequence. The numbers do not increase at a steady rate. There is a pattern, however: each number in the sequence is twice as large as the number before it. We'll look at this type of series shortly.

PROBLEM 14-3

Is the following sequence an arithmetic sequence? If so, what are the values s_0 (the starting value) and c (the constant of change)?

$$S = 100, 135, 170, 205, 240, 275, 310, \ldots$$

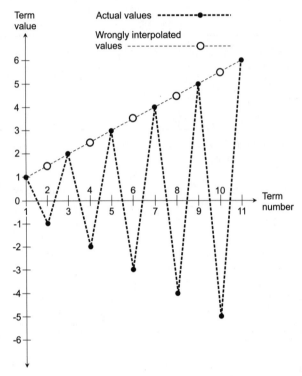

Fig. 14-2.   Graphical portrayal of mistaken interpolation.

SOLUTION 14-3
This is an arithmetic sequence, at least for the numbers shown (the first seven terms). In this case, $s_0 = 100$ and $c = 35$.

Growth by Multiplication

Another type of progression has values that are repeatedly multiplied by some constant. Here are a few examples:

$$G = 1, 2, 4, 8, 16, 32$$
$$H = 1, -1, 1, -1, 1, -1, \ldots$$
$$I = 1, 10, 100, 1000$$
$$J = -5, -15, -45, -135, -405$$
$$K = 3, 9, 27, 81, 243, 729, 2187, \ldots$$
$$L = 1/2, 1/4, 1/8, 1/16, 1/32, \ldots$$

Sequences *G*, *I*, and *J* are finite. Sequences *H*, *K*, and *L* are infinite, as indicated by the three dots following the last term in each sequence.

GEOMETRIC PROGRESSION

Examine the six sequences above. Upon casual observation, they appear to be much different from one another. But in all six sequences, each term is a specific and constant multiple of the term before it. Note:

- The values in *G* progress by a constant factor of 2.
- The values in *H* progress by a constant factor of –1.
- The values in *I* progress by a constant factor of 10.
- The values in *J* progress by a constant factor of 3.
- The values in *K* progress by a constant factor of 3.
- The values in *L* progress by a constant factor of 1/2.

Each sequence has a starting point or first number. After that, succeeding numbers are generated by repeated multiplication by a constant. If the constant is positive, the values in the sequence stay "on the same side of 0" (they either remain positive or remain negative). If the constant is negative, the values in the sequence "alternate to either side of 0" (if a given term is positive, the next is negative, and if a given term is negative, the next is positive).

Let t_0 be the first number in a sequence *T*, and let *k* be a constant. Imagine that *T* can be written in this form:

$$T = t_0, t_0 k, t_0 k^2, t_0 k^3, t_0 k^4, \ldots$$

for as long as the sequence goes. Such a sequence is called a *geometric sequence* or a *geometric progression*.

If *k* happens to be equal to 1, the sequence consists of the same number, listed over and over. If $k = -1$, the sequence alternates between t_0 and its negative. If t_0 is less than −1 or greater than 1, the values get farther and farther from 0. If t_0 is between (but not including) −1 and 1, the values get closer and closer to 0. If $t_0 = 1$ or $t_0 = -1$, the values stay the same distance from 0.

The numbers t_0 and *k* can be whole numbers, but this is not a requirement. As long as the multiplication factor between any two adjacent terms in a sequence is the same, the sequence is a geometric progression. In the last sequence, $k = 1/2$. This is an especially interesting sequence, as we'll see in a moment.

GEOMETRIC SERIES

For a geometric sequence, the corresponding *geometric series*, which is the sum of all the terms, can always be defined if the sequence is finite. Sometimes the sum of all the terms can be defined even if the sequence is infinite.

For the above sequences G through L, let the corresponding series be called G_+ through L_+. Then:

$$G_+ = 1 + 2 + 4 + 8 + 16 + 32 = 63$$
$$H_+ = 1 - 1 + 1 - 1 + 1 - 1 + \ldots = ?$$
$$I_+ = 1 + 10 + 100 + 1000 = 1111$$
$$J_+ = -5 - 15 - 45 - 135 - 405 = -605$$
$$K_+ = \text{``blows up'' and is not defined}$$
$$L_+ = 1/2 + 1/4 + 1/8 + 1/16 + 1/32 + \ldots = ?$$

The finite series G_+, I_+, and J_+ are straightforward enough. The infinite series H_+ seems unable to settle on 0 or 1, repeatedly hitting both. It's tempting to say that H_+ is a number with two values at once, and a fascinating theory can be built around the notion of multi-valued numbers. But in conventional math, we have to say that H_+ is not definable as a number. The infinite series K_+ runs off "out of control" and is an example of a *divergent series*, because its values keep on getting farther and farther away from 0 without limit.

That leaves us with L_+. What's going on with this series?

PARTIAL SUMS

When we have an infinite sequence and we start to add up its numbers, we get another sequence of numbers representing the sums. These sums are called *partial sums*. For the above sequences H, K, and L, the partial sums, which we will denote using asterisk superscripts instead of plus-sign subscripts, go like this:

$$H = 1, -1, 1, -1, 1, -1, \ldots$$
$$H^* = 1, 0, 1, 0, 1, 0, \ldots$$
$$K = 3, 9, 27, 81, 243, 729, 2187, \ldots$$
$$K^* = 3, 12, 39, 120, 363, 1092, 3279, \ldots$$
$$L = 1/2, 1/4, 1/8, 1/16, 1/32, \ldots$$
$$L^* = 1/2, 3/4, 7/8, 15/16, 31/32, \ldots$$

The partial sums denoted by $H*$ and $K*$ don't settle down on anything. But the partial sums denoted by $L*$ seem to approach 1. They aren't skyrocketing off into uncharted territory, and they aren't alternating between or among any multiple numbers. They are under control. The partial sums in $L*$ seem to have a clear destination that they could reach, if only they had an infinite amount of time to get there.

CONVERGENT SERIES

It turns out that L_+, representing the sum of the infinite string of numbers L, adds up to exactly 1! We can demonstrate this by observing that the partial sums "close in" on a value of 1. As the position in the sequence of partial sums, $L*$, gets farther and farther along, the denominators keep doubling, and the numerator is always 1 less than the denominator. In fact, if we want to find the nth number L_n in the sequence of partial sums $L*$, we can calculate it by using the following formula:

$$L_n = (2^n - 1)/2^n$$

As n becomes large, 2^n becomes large much faster, and the proportional difference between $2^n - 1$ and 2^n becomes smaller and smaller. When n is extremely large, the quotient $(2^n - 1)/2^n$ is almost exactly equal to 1. We can make the quotient as close to 1 as we want by going out far enough in the series of partial sums, but we can never make it bigger than 1. The series L_+ is said to *converge* on the number 1, and as such it is an example of a *convergent series*.

GEOMETRIC INTERPOLATION

When you see a long sequence of numbers, you can usually figure out, without much effort, whether or not it's a geometric sequence. If it isn't immediately obvious, you can conduct a test: divide each number by the one before it. If all the quotients are the same, then the sequence is a geometric sequence.

Now imagine, as you did with the arithmetic sequences earlier in this chapter, that you're given a long string of numbers, but some of them are missing. An example of such a situation is shown in Table 14-4. It's easy to determine the missing values, once you notice that this is a geometric sequence in which each term is twice as big as the term preceding it. The 4th term has a value of 8, and the 9th term has a value of 256. We can call the process of filling in Table 14-4 an example of *geometric interpolation*.

Table 14-4   Another example of a sequence with some intermediate values missing. The missing values can be filled in by interpolation.

Position:	1st	2nd	3rd	4th	5th	6th	7th	8th	9th	10th	11th
Value:	1	2	4	?	16	32	64	128	?	512	1024

Table 14-5   Extension of a geometric sequence can be done by extrapolation.

Position:	1st	2nd	3rd	4th	5th	6th	7th	8th	9th	10th	11th
Value:	32	16	8	4	2	1	1/2	?	?	?	?

GEOMETRIC EXTRAPOLATION

Just as a geometric sequence can be interpolated, it's also possible to "predict" what the next numbers are, or would be if the series were lengthened. Table 14-5 illustrates an example of this type of situation. The values in this sequence are repeatedly cut in half. The 7th number is 1/2. Therefore, the 8th number must be 1/4, the 9th number must be 1/8, the 10th number must be 1/16, and the 11th number must be 1/32. We can call the process of filling in Table 14-5 *geometric extrapolation.*

PLOTTING A GEOMETRIC SEQUENCE

A geometric sequence, like an arithmetic sequence, looks like a set of points when plotted on a Cartesian plane. Figure 14-3 shows examples of two geometric sequences as they appear when graphed. (The dashed curves aren't actually parts of the sequences.) Note that the dashed curves are not straight, but they are smooth.

One of the sequences in Fig. 14-3 is increasing, and the dashed curve connecting this set of points goes upward as you move to the right. Because this sequence is finite, the dashed curve ends. The other sequence is decreasing, and the dashed curve goes downward and approaches 0 as you move to the right. This sequence is infinite, as shown by the three dots at the end of the string of numbers, and also by the arrow at the right-hand end of the dashed curve.

If a geometric sequence has a negative factor, that is, if $k < 0$, the plot of the points alternates back and forth on either side of 0. In this case, the points fall along two different curves, one above the horizontal axis and

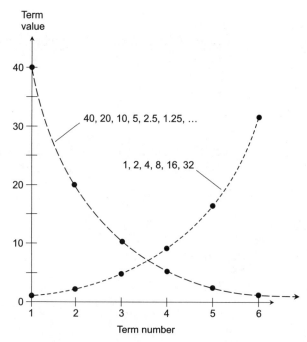

Fig. 14-3.   Plots of values of the terms in two different geometric sequences.

the other below. If you want to see what happens in a case like this, try plotting an example. Try setting $t_0 = 64$ and $k = -1/2$, and plot the resulting points.

PROBLEM 14-4

Suppose you get a certificate of deposit (CD) at your local bank for $1000.00, and it earns interest at the annualized rate of exactly 5% per year. How much will this CD be worth at the end of 6 years, assuming a constant interest rate? Round all figures off to the nearest cent at the end of every year.

SOLUTION 14-4

The CD will be worth $1340.10 after 6 years. To calculate this, multiply $1000 by 1.05, then multiply this result by 1.05, and repeat this process a total of six times. The resulting numbers are a geometric progression.

- After 1 year: $1000.00 × 1.05 = $1050.00
- After 2 years: $1050.00 × 1.05 = $1102.50
- After 3 years: $1102.50 × 1.05 = $1157.63
- After 4 years: $1157.63 × 1.05 = $1215.51
- After 5 years: $1215.51 × 1.05 = $1276.29
- After 6 years: $1276.29 × 1.05 = $1340.10

PROBLEM 14-5

Is the following sequence a geometric sequence? If so, what are the values t_0 (the starting value) and k (the factor of change)?

$$T = 3, -6, 12, -24, 48, -96, \ldots$$

SOLUTION 14-5

This is a geometric sequence. The numbers change by a factor of -2. Each number in the sequence is the product of -2 and the number before it. In this case, $t_0 = 3$ and $k = -2$.

PROBLEM 14-6

Is the following sequence a geometric sequence, an arithmetic sequence, or neither? Is there any pattern to it? If so, what is the pattern?

$$T = 10, 13, 17, 22, 28, 35, 43, \ldots$$

SOLUTION 14-6

This is not a geometric sequence. It isn't an arithmetic sequence either. But there is a pattern. The difference between the first and second numbers is 3, the difference between the second and third numbers is 4, the difference between the third and fourth numbers is 5, and so on; the difference keeps increasing by 1 for each succeeding pair of numbers. (Some sequences have patterns that are subtle indeed. Computers can be helpful in analyzing such sequences.)

PROBLEM 14-7

Suppose a particular species of cell undergoes *mitosis* (splits in two) without fail every half hour, precisely on the half hour. We take our first look at a petri dish at 12:59 p.m., and find 3 cells. At 1:00 p.m., mitosis occurs for all the cells at the same time, and then there are 6 cells in the petri dish. At 1:30 p.m., mitosis occurs again, and then there are 12 cells. How many cells are there in the petri dish at 4:01 p.m.?

SOLUTION 14-7

There are 3 hours and 2 minutes between 12:59 p.m. and 4:01 p.m. This means that mitosis takes place 7 times: at 1:00, 1:30, 2:00, 2:30, 3:00, 3:30, and 4:00. Table 14-6 illustrates the scenario. There are 384 cells at 4:01 p.m., just after the 7th mitosis event that occurs at 4:00.

Table 14-6   Table for Problem 14-7.

Time:	12:59	1:01	1:31	2:01	2:31	3:01	3:31	4:01
Number of cells:	3	6	12	24	48	96	192	384

Exponential Functions

The idea of the geometric progression, in which a value is repeatedly multiplied by some constant, can be extended into the general realm of continuous-curve functions. In an *exponential function*, a constant is raised to some variable power.

VARIABLE EXPONENTS

Until now, the notion of "raising something to a power" has been kept simple because we've dealt only with whole-number exponents. For example, when a variable x is raised to the power of 5, we write it as x^5. When some constant k is raised to the power of 10, we write it as k^{10}. But what about exponents that are not whole numbers? What, for example, is meant by the expression $x^{1.5}$ or $7^{2/3}$ or $z^{-8/5}$? What if an exponent is an irrational number such as π, the ratio of a circle's circumference to its diameter?

All scientific calculators, and even some ordinary ones, have a key marked something like "x^y" or "$x\hat{\ }y$." This is literally read as, "x to the yth power." To find the value of, say, $3^{1.5}$, you first enter the number 3, then you hit the "x^y" or "$x\hat{\ }y$" key, and finally you enter 1.5. This should give you a result of 5.19615.... To find the value of $3^{2.88}$, you first enter the number 3, then you hit the "x^y" or "$x\hat{\ }y$" key, and finally you enter 2.88. This will give you 23.6651.... You can even find the value of 3 raised to some power between 0 and 1, or less than 0 (that is, negative).

The general term *exponential* refers to the raising of a constant to some power, where that power can be any sort of number. An exponential function is the raising of a constant to a variable power. Suppose the following relationship exists among three real numbers a, x, and y:

$$y = a^x$$

In this case, x is the independent variable and y is the dependent variable. The constant a is called the *base*.

The two most common exponential-function bases are 10 and the *natural exponential base*. The natural exponential base is symbolized by the lower-case, italicized letter e, and is an irrational number. It is approximately equal to 2.71828.

COMMON EXPONENTIALS

Base-10 exponentials are also known as *common exponentials*. For example:

$$10^{-3.000} = 0.001$$

Figure 14-4 is an approximate graph of the function $y = 10^x$. The domain of this function encompasses the entire set of real numbers. The range is limited to positive real numbers only.

NATURAL EXPONENTIALS

Base-e exponentials are also known as *natural exponentials*. For example:

$$e^{-3.000} \approx 2.71828^{-3.000} \approx 0.04979$$

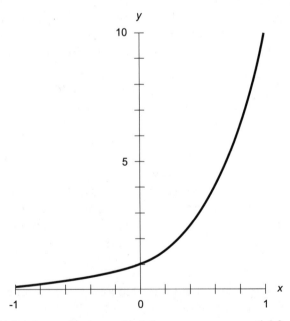

Fig. 14-4.   Approximate graph of the common exponential function.

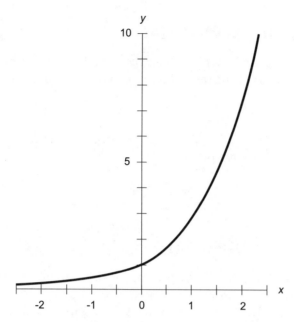

Fig. 14-5.   Approximate graph of the natural exponential function.

Figure 14-5 is an approximate graph of the function $y = e^x$. The domain encompasses the entire set of real numbers. The range is limited to positive real numbers only.

HOW DO WE FIND THEM?

Common and natural exponentials are easy to find. All you have to do is get, and use, a scientific calculator. The calculator can show us only the first few digits of the result, so the answer is usually not exact, but it's a good enough approximation for most purposes.

Rules for Exponentials

Exponential functions allow us to manipulate numbers in ways that can be useful in the sorts of work you're likely to encounter these days. Here are some of the most common "laws" concerning exponentials. These laws, like the basic rules of arithmetic, are worth memorizing. That idea is repulsive to some folks, but it's the truth. It is good to know the following rules by heart.

RECIPROCAL OF EXPONENTIAL

Suppose that x is a real number. The reciprocal of the common (base-10) exponential of x is equal to the common exponential of the negative of x:

$$1/(10^x) = 10^{-x}$$

The reciprocal of the natural (base-e) exponential of x is equal to the natural exponential of the negative of x:

$$1/(e^x) = e^{-x}$$

PRODUCT OF EXPONENTIALS

Let x and y be real numbers. The product of the common exponentials of x and y is equal to the common exponential of the sum of x and y. You might recognize this rule from all the way back in Chapter 3, when we worked with power-of-10 notation. When two numbers in scientific notation are multiplied, the powers of 10 add:

$$(10^x)(10^y) = 10^{(x+y)}$$

The product of the natural exponentials of x and y is equal to the natural exponential of the sum of x and y:

$$(e^x)(e^y) = e^{(x+y)}$$

RATIO OF EXPONENTIALS

Let x and y be real numbers. The ratio (quotient) of the common exponentials of x and y is equal to the common exponential of the difference between x and y. This rule, too, appeared in Chapter 3. When two extreme numbers are divided in scientific notation, their powers of 10 subtract:

$$10^x/10^y = 10^{(x-y)}$$

The ratio of the natural exponentials of x and y is equal to the natural exponential of the difference between x and y:

$$e^x/e^y = e^{(x-y)}$$

EXPONENTIAL OF EXPONENTIAL

Let x and y be real numbers. The yth power of the quantity 10^x is equal to the common exponential of the product xy:

$$(10^x)^y = 10^{(xy)}$$

The yth power of the quantity e^x is equal to the natural exponential of the product xy:

$$(e^x)^y = e^{(xy)}$$

PROBLEM 14-8

Suppose a scientist, Professor P, has developed a theory to the effect that the world population increases exponentially with time. Imagine that the present population of the world has been accurately determined, and we assign this number (whatever it happens to be) the value 100%, or "exactly 1 population unit." Professor P then tells us that every 100 years, the population increases by precisely a factor of 10, known as an *order of magnitude*. Suppose the theory of Professor P is correct, and that the theory remains valid for millennia to come. What will be the world's "people count," in population units, 100 years from now? What will it be in 200 years? In 300 years?

SOLUTION 14-8

In 100 years (1 century from now), the world's "people count" will be 10^1, or 10, population units. In 200 years (2 centuries), it will be 10^2, or 100, population units. In 300 years, it will be 10^3, or 1000, population units. This means that in a century, the world will contain 10 times as many people as it does now; in 2 centuries the world will have 100 times as many people as it does now; in 3 centuries there will be 1000 times as many people on the planet as there are today. Let's hope that Professor P's theory doesn't hold true that long!

PROBLEM 14-9

Write down an exponential equation showing the world population p, in population units, as a function of time t, in centuries, according to the theory of Professor P. Assume that $t=0$ right now.

SOLUTION 14-9

The equation looks like this:

$$p = 10^t$$

You can check this out by plugging in values for t. If $t=0$, then $p=10^0=1$. If $t=1$, then $p=10^1=10$. If $t=2$, then $p=10^2=100$. If $t=3$, then $p=10^3=1000$. These results agree with those obtained in Solution 14-8.

PROBLEM 14-10
According to the theory of Professor P, how many people, in population units, will inhabit the world 150 years from now?

SOLUTION 14-10
To solve this problem, we use the equation from Solution 14-9, and plug in the value $t=1.5$. This gives us a population p as follows:

$$p = 10^{1.5}$$

How do we calculate the value of an expression like $10^{1.5}$? It's easy if you have a scientific calculator that includes exponential functions. Use the "x^y" or "$x^{\wedge}y$" key. Your calculator might also have a "10^x" key. In any event, you should obtain the value $p=31.62$, which we can round off to 32. This means that in 150 years, according to the theory of Professor P, the world population will be 32 times as great as it is right now.

Logarithms

A *logarithm* (sometimes called a *log*) is an exponent to which a constant is raised to obtain a given number. Suppose the following relationship exists among three real numbers a, x, and y:

$$a^y = x$$

Then y is the *base-a logarithm* of x. This expression is written

$$y = \log_a x$$

The two most common logarithm bases are 10 and e, the same as the exponential bases.

COMMON LOGARITHMS

Base-10 logarithms are also known as *common logarithms* or *common logs*. In equations, common logarithms are denoted by writing "log" without a subscript, or occasionally "log" with a subscript 10. For example:

$$\log 100 = \log_{10} 100 = 2.00$$

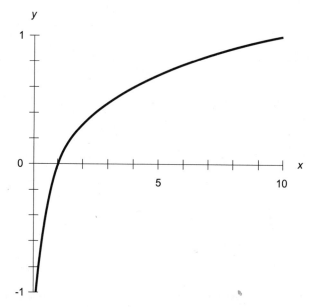

Fig. 14-6.   Approximate graph of the common logarithm function.

Figure 14-6 is an approximate graph of the function $y = \log x$. The domain is limited to the positive real numbers. The range encompasses the entire set of real numbers.

NATURAL LOGARITHMS

Base-e logarithms are also called *natural logs* or *Napierian logs*. In equations, the natural-log function is usually denoted "ln" or "$\log_e$." For example:

$$\ln 2.71828 = \log_e 2.71828 \approx 1.00000$$

Figure 14-7 is an approximate graph of the function $y = \ln x$. The domain is limited to the positive real numbers, just as in the case of the common log function. The range of the natural log function, also like the common log function, encompasses the entire set of real numbers.

HOW DO WE FIND THEM?

Common and natural logarithms can be easily and quickly found using a scientific calculator. That's all you have to know for practical purposes.

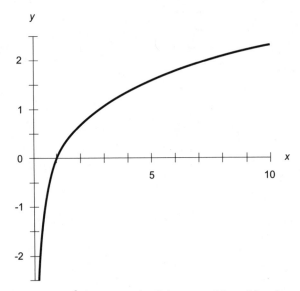

Fig. 14-7. Approximate graph of the natural logarithm function.

Once you have learned which keys to use (the instructions can help you with this if it isn't obvious), finding logarithms is painless.

Rules for Logarithms

Logarithmic functions, like their exponential-function counterparts, make it possible to work with numbers in unique ways. And, as with exponentials, logarithms have a way of coming up in work nowadays. Here are some rules you should know about logarithmic functions.

LOGARITHM OF PRODUCT

If x and y are both positive real numbers, the common logarithm of the product is equal to the sum of the common logarithms of the individual numbers:

$$\log xy = \log x + \log y$$

The natural logarithm of the product is equal to the sum of the natural logarithms of the individual numbers:

$$\ln xy = \ln x + \ln y$$

LOGARITHM OF RATIO

Let x and y be positive real numbers. The common logarithm of their ratio, or quotient, is equal to the difference between the common logarithms of the individual numbers:

$$\log(x/y) = \log x - \log y$$

The natural logarithm of their ratio, or quotient, is equal to the difference between the natural logarithms of the individual numbers:

$$\ln(x/y) = \ln x - \ln y$$

LOGARITHM OF QUANTITY RAISED TO A POWER

Let y be a real number, and let x be a positive real number. The common logarithm of x raised to the power y can be expressed as a product:

$$\log x^y = y \log x$$

The natural logarithm of x raised to the power y can be expressed as a product:

$$\ln x^y = y \ln x$$

LOGARITHM OF RECIPROCAL

Let x be a positive real number. The common logarithm of the reciprocal of x is equal to the negative of the common logarithm of x, as follows:

$$\log(1/x) = -\log x$$

The natural logarithm of the reciprocal of x is equal to the negative of the natural logarithm of x, as follows:

$$\ln(1/x) = -\ln x$$

LOGARITHM OF THE ROOT OF A QUANTITY

Let x be a positive real number. Let y be any real number except 0. The common logarithm of the yth root of x (also denoted as x to the $1/y$ power) is given by:

$$\log x^{(1/y)} = (\log x)/y$$

The natural logarithm of the yth root of x (also denoted as x to the $1/y$th power) is given by:

$$\ln x^{(1/y)} = (\ln x)/y$$

PERCEIVED INTENSITY

If you have ever done serious work with high-fidelity (hi-fi) equipment, you've heard about units called *decibels*, symbolized dB. The decibel is a logarithmic unit. It is used to express the relative intensity of sound waves, radio signals, and even visible light.

The concept of the decibel evolved because our senses perceive variable effects according to the logarithm of the intensity, not in direct proportion to the intensity. This is a defense mechanism against the harshness of the world. If we perceived phenomena such as sound and light in direct proportion to their true levels, the variability would be overwhelming. We might not be able to hear someone whispering a meter away, yet a trumpet blast from across a room would knock us out. We might not be able to see the full moon on a clear night, but we would be blinded by the light of a cloudy day.

Imagine two sounds having the same frequency and wave shape, such as middle C played on a trombone. Call them "sound number 1" and "sound number 2." Imagine that a trombone plays note 1, pauses for a moment, and then plays note 2. Notes 1 and 2 differ in intensity by 1 dB if the change is just barely enough so that a listener can tell the difference *when the change is expected*. This turns out to be a difference in audio power of approximately 26%, or a ratio of 1.26:1. If the change is not expected, the smallest detectable audio power difference is about 100%, or a ratio of 2:1 (believe it or not).

If P_1 is the power (in watts) contained in sound 1 and P_2 is the power (also in watts) contained in sound 2, their relationship in decibels, R, is given by this formula:

$$R = 10 \log(P_2/P_1)$$

This formula also works for radio-signal power and light-brilliance power. If we know the relationship R in decibels between phenomenon number 1 and phenomenon number 2, then the actual power ratio P_2/P_1 is a common exponential function of R, as follows:

$$P_2/P_1 = 10^{(R/10)}$$

PROBLEM 14-11
Suppose two sounds differ in volume by a power ratio of 2:1, which allegedly
is the smallest difference a listener can detect if the change is not anticipated.
What is the ratio in decibels?

SOLUTION 14-11
Use the formula above. Let $P_2 = 2$ and $P_1 = 1$. Then:

$$R = 10\log(P_2/P_1)$$
$$= 10\log(2/1)$$
$$= 10\log 2$$
$$= 10 \times 0.30103$$
$$= 3.0103 \text{ dB}$$

We can round this off to 3 dB.

PROBLEM 14-12
Suppose you are listening to a musical recording and the needles on your hi-fi
amplifier are peaking at a reading of 0 dB. You boost the volume until the
needles kick up to +10 dB. By what factor have you increased the audio
power level?

SOLUTION 14-12
In the formula, set $R = 10$ because the change in sound volume is +10 dB.
From this, calculate the audio power ratio:

$$P_2/P_1 = 10^{(R/10)}$$
$$= 10^{(10/10)}$$
$$= 10^1$$
$$= 10$$

This means you have increased the audio power by a factor of 10:1.

Graphs Based on Logarithms

Logarithms make it possible to graph certain functions that don't lend themselves to clear portrayal on the rectangular (Cartesian) coordinate plane.
This is done by making the increments on one or both axes proportional
to the logarithm of the variable value, rather than directly proportional to

the variable value. A scale or axis in which the sizes of the increments are in direct proportion to the variable value is called a *linear scale* or a *linear axis*. A scale or axis in which the sizes of the increments are in proportion to the logarithm of the variable value is called a *logarithmic scale* or a *logarithmic axis*.

SEMILOG (*x*-LINEAR) COORDINATES

Figure 14-8 shows *semilogarithmic* (*semilog*) *coordinates* for defining points in a portion of the *xy*-plane. The independent-variable (*x*) axis is linear, and the dependent-variable (*y*) axis is logarithmic. In this example, functions can be plotted with domains and ranges as follows:

$$-1 \le x \le 1$$
$$0.1 \le y \le 10$$

The *y* axis in Fig. 14-8 spans two orders of magnitude (powers of 10). The span could be larger or smaller than this, but in any case the *y* values cannot extend all the way down to 0. This is because the logarithm of 0 is undefined.

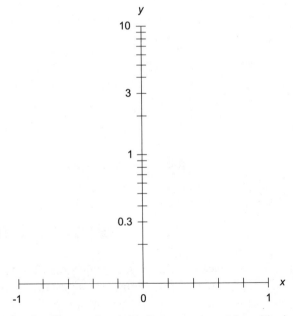

Fig. 14-8.   Semilog *xy*-plane with linear *x* axis and logarithmic *y* axis.

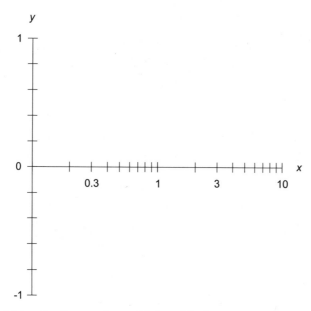

Fig. 14-9. Semilog xy-plane with logarithmic x axis and linear y axis.

SEMILOG (y-LINEAR) COORDINATES

Figure 14-9 shows a different sort of semilog coordinate system for defining points in a portion of the xy-plane. Here, the independent-variable (x) axis is logarithmic, and the dependent-variable (y) axis is linear. In this example, functions can be plotted with domains and ranges as follows:

$$0.1 \leq x \leq 10$$
$$-1 \leq y \leq 1$$

The x axis in Fig. 14-9 spans two orders of magnitude. The span could be larger or smaller, but in any case the x values cannot extend all the way down to 0.

LOG–LOG COORDINATES

Figure 14-10 shows *log–log coordinates* for defining points in a portion of the xy-plane. Both axes are logarithmic. In this example, functions can be plotted with domains and ranges as follows:

$$0.1 \leq x \leq 10$$
$$0.1 \leq y \leq 10$$

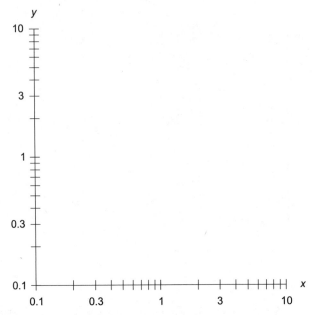

Fig. 14-10. Log–log xy-plane. Both axes are logarithmic.

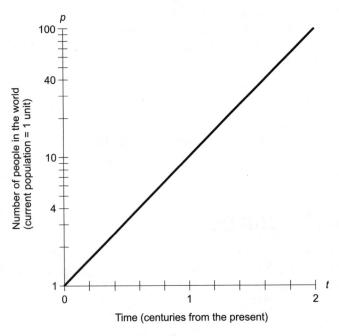

Fig. 14-11. Illustration for Problem 14-13.

The axes in Fig. 14-10 span two orders of magnitude. The span of either axis could be larger or smaller, but in any case the values cannot extend all the way down to 0.

PROBLEM 14-13
In semilog coordinates, graph the world population p, in population units, as a function of time t, in centuries, according to the theory of Professor P. Show the function over the domain from $t=0$ (right now) to $t=2$ (2 centuries, or 200 years, from now). Use a linear scale for the horizontal axis and a logarithmic scale for the vertical axis.

SOLUTION 14-13
Refer to Fig. 14-11. Note that the graph is a straight line. One of the major assets of semilogarithmic coordinates is the fact that they can show exponential and logarithmic functions as straight lines, rather than as curves.

Quiz

Refer to the text in this chapter if necessary. A good score is eight correct. Answers are in the back of the book.

1. Imagine a frog that jumps halfway to a wall. After 10 seconds, it jumps halfway to the wall again. After another 10 seconds, it jumps halfway to the wall once again. It keeps repeating this process. In theory, how long will it take to reach the wall after its first jump?
    (a) More information is needed to answer this question.
    (b) 10^2, or 100, seconds.
    (c) 2^{10}, or 1024, seconds.
    (d) It will never reach the wall.

2. Imagine a frog that jumps halfway to a wall. After 4 seconds, it jumps halfway to the wall again. After 2 seconds, it jumps halfway to the wall again. After 1 second, it jumps halfway to the wall again. The frog keeps repeating this process, but the intervals between jumps keep getting half as long: 1/2 second, then 1/4 second, and so on. In theory, how long will it take to reach the wall, if we start timing it the moment it finishes its first jump?
    (a) More information is needed to answer this question.
    (b) 8 seconds.

(c) 4^2, or 16, seconds.
(d) 2^8, or 256, seconds.

3. Consider this sum of numbers:

$$S = 4 - 4 + 4 - 4 + 4 - 4 + 4 - 4 + ...$$

Which of the following statements is true?
(a) S is a convergent series.
(b) S is an arithmetic series.
(c) S is an infinite series.
(d) S is not a series.

4. What is the base-10 logarithm of 0.00001?
(a) $e/10$
(b) $-e/10$
(c) 5
(d) -5

5. Which of the following points cannot, in theory, be shown on a log–log (x,y) coordinate system?
(a) The point corresponding to $x=1$ and $y=1$.
(b) The point corresponding to $x=0$ and $y=1$.
(c) The point corresponding to $x=10$ and $y=10$.
(d) The point corresponding to $x=0.1$ and $y=0.1$.

6. What is the natural logarithm of e?
(a) 0
(b) 1
(c) e
(d) $1/e$

7. Suppose a sequence of numbers is such that each value is exactly 4 less than the value before it. This is an example of
(a) a geometric progression
(b) an arithmetic progression
(c) a logarithmic progression
(d) a fractional progression

8. Suppose you're listening to a compact disc on your new high-powered stereo sound system. The VU (volume unit) meter needles are kicking up to the points marked 0 dB, exactly where the black lines end and the red lines begin. If you increase the gain so the needles are kicking up to +6 dB in the red zone, the sound power coming from

the speakers has approximately
(a) doubled
(b) quadrupled
(c) increased by a factor of 6
(d) increased by a factor of 2^6, or 64

9. Suppose you take a 4-week temporary job with a 5-day work week. The boss promises that you'll be paid daily in US dollars ($): $1.00 at the end of the first day, $2.00 at the end of the second day, and doubling each day after that for the entire period. What should this indicate to you?
    (a) The boss is a liar or an unwise wealthy person.
    (b) The payment is meager, and you should look for another job.
    (c) Your accumulated pay, day by day, will represent a convergent series.
    (d) Nothing, without more information.

10. Suppose a sequence of numbers is such that each value is exactly 25% of the value before it. This is an example of
    (a) a geometric progression
    (b) an arithmetic progression
    (c) a logarithmic progression
    (d) a fractional progression

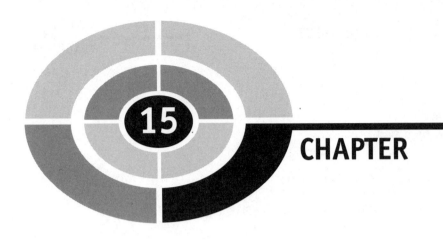

CHAPTER 15

How Things Move

What are you really doing when you drive a 2-ton vehicle down the Interstate at high speed? Why does your suburban utility vehicle try to go off the road if you take a curve too fast? Why should you never drive a pickup truck full of loose, heavy bricks? In this chapter, we'll take a close look at the "why and wherefore" behind how things move, and what they can do when they move.

Mass and Force

The term *mass* refers to sheer quantity of matter, in terms of its ability to resist motion when acted upon by a *force*. A good synonym for mass is *heft*. Every material object has a specific, definable mass. The sun has a certain mass; earth has a much smaller mass. A lead shot has a far smaller mass still. Even subatomic particles, such as protons and neutrons, have mass.

MASS IS A SCALAR

The mass of an object or particle has magnitude (size or extent), but not direction. The mass of any object can be quantified in units such as kilograms (kg). Mass is customarily denoted by the lowercase italic letter m.

You might think that mass can have direction. When you stand somewhere, your body presses downward on the floor or the pavement or the ground. If someone is more massive than you, his body presses downward too, but harder. If you get in a car and accelerate, your body presses backward in the seat as well as downward toward the center of the earth. But this pressing-down or pressing-back is force, not mass. The force you feel is caused in part by your mass, and in part by gravity or acceleration. Mass itself has no direction. It's like temperature or sound intensity. Mass is a *scalar quantity* because it can be expressed by a plain, ordinary number.

HOW MASS IS DETERMINED

The simplest way to determine the mass of an object is to measure it with a scale. But this isn't the best way. When you put something on a scale, you are measuring that object's *weight* in the gravitational field of the earth. This field has about the same intensity, no matter where you go on the planet. But it's different on the moon or on other planets. The mass of a 1-kg bag of dried peas is the same wherever you take it on earth. But it will weigh less on the moon, because the moon's gravitational field is weaker.

Let's conduct a little thought experiment. Suppose you are on an interplanetary journey, coasting along on your way to Mars, and everything in your space vessel is weightless. How can you measure the mass of an object, such as a lead shot, under these conditions? It floats around in the cabin along with your body, the pencils you write with, and everything else that is not tied down. You are aware that the lead shot is more massive than, say, a pea, but how can you measure it to be certain?

One way to measure mass, independently of gravity, involves using a pair of springs set in a frame, with the object placed in the middle (Fig. 15-1). If you put something between the springs and pull it to one side, the object oscillates, or "see-saws." You try this with a pea, and it "see-saws" rapidly. You try it again with a lead shot, and the springs oscillate slowly. This *mass meter* is anchored to a wall in the space ship's cabin. (Anchoring the mass meter keeps it from wagging back and forth in mid-air after you start an object oscillating against the springs.)

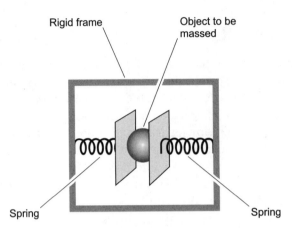

Rigid frame   Object to be
                     massed

Spring   Spring

Fig. 15-1.   Mass can be measured by getting an object to "see-saw" between a pair of springs
in a weightless environment.

A scale of this type must be calibrated in advance before it can render meaningful figures for masses of objects. The calibration can be shown as a graph of the *oscillation period* (the time it takes for the object to complete one cycle of "see-saw" motion) or *oscillation frequency* (the number of complete "see-saw" cycles per second) as a function of the mass. Once this calibration is done in a weightless environment and the graph has been drawn, you can use the spring device and the graph to measure the mass of anything within reason. The readings will be thrown off if you try to use the mass meter on earth, on the moon, or on Mars, because there is an outside force – gravity – acting on the mass. The same problem will occur if you try to use the scale when the space ship is accelerating, rather than merely coasting or orbiting through space.

PROBLEM 15-1
Suppose you place an object in a mass meter similar to the one shown in Fig. 15-1. Also suppose that the mass-versus-frequency calibration graph looks like Fig. 15-2. The object "see-saws" at a rate of 5 oscillations per second (that is, a frequency of 5 Hz). What is the approximate mass of this object?

SOLUTION 15-1
Locate the frequency on the horizontal scale. Draw a vertical line (or place a ruler) parallel to the vertical (mass) axis. Note where this straight line intersects the curve. Draw a horizontal line from this intersection point toward the left until it hits the mass scale. Read the mass off the scale. It is approximately 0.8 kg, as shown in Fig. 15-3.

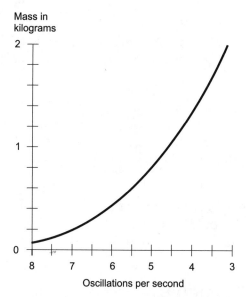

Fig. 15-2.   Graph of mass versus oscillation ("see-sawing") frequency for a spring-type mass meter.

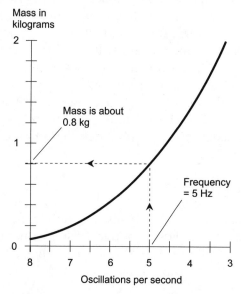

Fig. 15-3.   Solution to Problem 15-1.

PROBLEM 15-2

What will the mass meter described in Problem 15-1 do if a mass of only 10 grams (10 g) is placed in between the springs?

SOLUTION 15-2
The scale will oscillate much faster. We don't know exactly how fast, because 10 g is only 0.01 kg, which is too small to show up on the graph in Fig. 15-2.

PROBLEM 15-3
Wouldn't it be easier, and more accurate in real life, to program the calibration curve for a mass meter into a computer, instead of using graphs like the ones shown here? That way, we could simply input the "see-saw" frequency into the computer, and read the mass on the display.

SOLUTION 15-3
Yes, that would be easier. In a real-life mass meter of the sort we've been discussing here, that would be done. In fact, we should expect such a scale to have a built-in microcomputer and display to tell us the mass directly.

BRICK VERSUS MARBLE

Imagine again that you are in a spacecraft, which is coasting along so everything in the cabin is weightless. Two objects float in front of you: a brick and a marble. Either the brick or the marble can be made to move across the cabin if you give it a push. But you know that the brick is more massive.

Suppose you flick your finger against the marble. It flies across the cabin and bounces off the wall. Then you flick your finger just as hard (no more, no less) against the brick. The brick takes several minutes to float across the cabin and bump into the opposite wall. The flicking of your finger imparts a force to the marble or the brick for a moment. The force is the same in either case, but it has a different effect on the brick than on the marble.

FORCE AS A VECTOR

Force is a *vector quantity*. It can have any magnitude, from the flick of a finger to a swift leg kick, the explosion of powder in a cannon, or the thrust of a rocket engine. But in addition to magnitude, force has direction. Vectors are symbolized using boldface letters of the alphabet. A *force vector*, for example, can be denoted **F**.

Sometimes, the direction of a force is not important. In such instances, we can speak of the magnitude of a force vector, and denote it as an italic letter such as F. The standard international unit of force magnitude is the *newton* (N), which is the equivalent of a *kilogram-meter per second squared* $(kg \cdot m/s^2)$. What does this mean? Here's an example: suppose the brick in

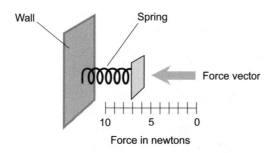

Fig. 15-4.   A device for measuring force.

your spacecraft has a mass of 1 kg, and you push against it with a force of 1 N for 1 s, and then let go. The brick will then be traveling at a speed of 1 m/s. It will have gone from stationary (with respect to its surroundings) to a speed of 1 m/s, which might seem rather slow unless it hits someone. If you hurl it with a force of 2 N for 1 s, it will end up moving twice as fast, that is, 2 m/s.

HOW FORCE IS DETERMINED

Force can be measured by the effect it has on an object with mass. It can also be measured by the extent to which it stretches an elastic object such as a spring. The spring type mass meter described above can be modified to make a *force meter* if one half of it is taken away and a calibrated scale is placed alongside (Fig. 15-4). This scale must be calibrated in a lab before the force meter is used.

Displacement

Displacement is almost the same thing as *distance*, but not quite. Displacement is defined along a straight line in a certain direction. But distance can be measured any which way. We might say that Minneapolis, Minnesota is displaced 100 km from Rochester, Minnesota "as the crow flies," or along a straight line. If you were to drive along US Route 52, however, the distance is closer to 120 km, because that highway does not follow a straight path from Rochester to Minneapolis.

DISPLACEMENT AS A VECTOR

Displacement is a vector quantity because it has magnitude and direction. A displacement vector is denoted by a lowercase boldface letter such as **d**.

Displacement magnitude (distance) is denoted by a lowercase italic letter such as *d*.

The displacement vector **d** of Minneapolis, expressed with respect to Rochester, is about 100 km in a northwesterly direction "as the crow flies." Expressed as a compass bearing, the direction of **d** is around 320°, measured clockwise from north. But if we speak about driving along Route 52, we can only talk about distance, because the road is not straight or level all the way. The distance of Minneapolis from Rochester, measured along Route 52, is a scalar, and is equal to roughly $d = 120$ km.

HOW DISPLACEMENT IS DETERMINED

Distance can be determined by mechanical measurement, or by inferring it with observations and calculations. In the case of a car or truck driving along Route 52, distance is measured with an *odometer* that counts the number of wheel rotations, and multiplies this by the circumference of the wheel. In other real-world scenarios, distance can be measured with a *meter stick*, by *triangulation*, or by more sophisticated means such as the use of the *Global Positioning System* (GPS).

The direction component of a displacement vector is determined by measuring one or more angles or coordinates relative to a reference axis. In the case of a local region on the earth's surface, direction can be found by specifying the *azimuth*, which is the compass bearing (the angle around the horizon clockwise relative to true north). That is the scheme used by hikers and back-packers. In three-dimensional space, *direction angles* are used. There are various ways in which these can be defined, but we don't need to get into that here.

Speed and Velocity

Speed is an expression of the rate at which an object moves, as seen or measured by some stationary observer. But we have to be careful about this. "Stationary" is a relative term. People sitting in an airliner might consider themselves to be "stationary," but that is not true with respect to the earth! If you stand on a street corner, you might think you are "stationary." With respect to the earth, you are; with respect to the passengers on an airliner flying high above, you aren't.

SPEED IS A SCALAR

The standard unit of speed is the meter per second (m/s). A car driving along Route 52 might have a cruise control that you can set at, say, 25 m/s. Then, assuming the cruise control works properly, you will be traveling, relative to the pavement, at a constant speed of 25 m/s. This will be true whether you are on a level straightaway, rounding a curve, cresting a hill, or passing the bottom of a valley. Speed can be expressed as a simple number, and the direction is not important. Speed is a scalar quantity. In this discussion, let's symbolize speed by the lowercase italic letter *v*.

Speed can, and almost always does, change with time. If you hit the brakes to avoid a deer crossing the road, your speed suddenly decreases. As you pass the deer, relieved to see it bounding off into a field unharmed, you pick up speed again.

Speed can be considered as an average over time, or as an instantaneous quantity. In the foregoing example, suppose you are driving at 25 m/s and then see a deer, put on the brakes, slow down to a minimum of 10 m/s, watch the deer run away, and then speed up to 25 m/s again, all in a time span of one minute. Your *average speed* over that minute might be 17 m/s. But your *instantaneous speed* varies from instant to instant, and is 17 m/s for only two instants (one as you slow down, the other as you speed back up).

HOW SPEED IS DETERMINED

In an automobile or truck, speed is determined by the same odometer that measures distance. But instead of simply counting up the number of wheel rotations from a given starting point, a *speedometer* counts the number of wheel rotations in a given period of time. When the tire circumference is known, the number of wheel rotations in a certain time interval can be translated directly into speed. (But you have to be sure the tires on your car are the right size for the odometer. Otherwise the device will not show the true speed.) In the USA, speed is measured in *miles per hour*. In the International System it is expressed in *kilometers per hour* or in *meters per second*.

Most car and truck speedometers respond almost immediately to a change in speed. These instruments measure the rotation rate of a car or truck axle by another method, similar to that used by the engine's *tachometer* (a device that measures revolutions per minute, or rpm). A real-life car or truck speedometer measures instantaneous speed, not average speed. If you want to know the average speed you have traveled during a certain period of time, you must measure the distance on the odometer, and then divide by the time elapsed.

In a given period of time t, if an object travels over a distance d at an average speed v_{avg}, then the following formulas apply. These are all arrangements of the same relationship among the three quantities:

$$d = v_{avg}t$$
$$v_{avg} = d/t$$
$$t = d/v_{avg}$$

PROBLEM 15-4

Look at the graph of Fig. 15-5. Curve A is a straight line. What is the instantaneous speed v_{inst} at $t = 5$ seconds?

SOLUTION 15-4

The speed shown by curve A is constant. You can tell because the curve is a straight line. The number of meters per second does not change. In 10 seconds, the object travels 20 meters; that's $20/10 = 2.0$ meters per second. Therefore, the speed at $t = 5$ seconds is $v_{inst} = 2.0$ meters per second.

PROBLEM 15-5

What is the average speed v_{avg} of the object denoted by curve A in Fig. 15-5, during the time span from $t = 3$ seconds to $t = 7$ seconds?

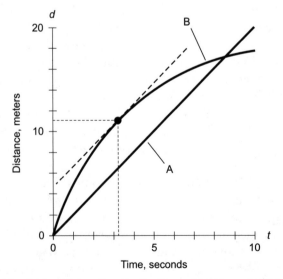

Fig. 15-5.   Illustration for Problems 15-4 through 15-8.

SOLUTION 15-5
Because the curve is a straight line, the speed is constant; we already know it is 2.0 meters per second. Therefore, $v_{avg} = 2.0$ meters per second between any two points in time shown in the graph.

PROBLEM 15-6
Examine curve B in Fig. 15-5. What can be said about the instantaneous speed of the object whose motion is described by this curve?

SOLUTION 15-6
The object starts out moving relatively fast, and the instantaneous speed decreases with the passage of time.

PROBLEM 15-7
Use visual approximation in the graph of Fig. 15-5. At what time t is the instantaneous speed v_{inst} of the object described by curve B equal to 2.0 meters per second?

SOLUTION 15-7
Use a straight-edge to visualize a line tangent to curve B whose slope is the same as that of curve A. That is, find the straight line, parallel to line A, that is tangent to curve B. Then locate the point on curve B where the line touches curve B. Finally, draw a line straight down, parallel to the distance (d) axis, until it intersects the time (t) axis. Read the value off the t axis. In this example, it appears to be approximately $t = 3.2$ seconds.

PROBLEM 15-8
Use visual approximation in the graph of Fig. 15-5. Consider the object whose motion is described by curve B. At the point in time t where the instantaneous speed v_{inst} is 2.0 meters per second, how far has the object traveled?

SOLUTION 15-8
Locate the same point that you found in Problem 15-7, corresponding to the tangent point of curve B and the line parallel to curve A. Draw a horizontal line to the left, until it intersects the distance (d) axis. Read the value off the d axis. In this example, it looks like it's about $d = 11$ meters.

VELOCITY IS A VECTOR

Velocity has two components: speed and direction. Thus, it is a vector quantity. You can't express velocity without defining both of these components. In the earlier example of a car driving along a highway from one town to

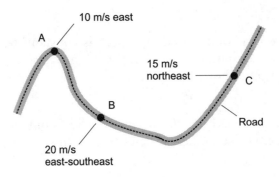

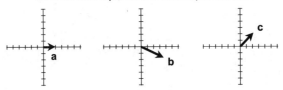

Fig. 15-6. Velocity vectors **a**, **b**, and **c** for a car at three points (A, B, and C) along a road.

another, the car's speed might be constant, but its velocity changes nevertheless. If you're moving along at a constant speed of 25 meters per second (25 m/s) and then you come to a bend in the road, your velocity changes because your direction changes.

Vectors can be graphically illustrated as line segments with arrowheads. The speed component of a velocity vector is denoted by the length of the line segment, and the direction is denoted by the orientation of the arrow. In Fig. 15-6, three velocity vectors are shown for a car traveling along a curving road. In this case, imagine the car moving generally from left to right. Three points are shown, called A, B, and C. The corresponding vectors are **a**, **b**, and **c**. Both the speed and the direction of the car change with time.

HOW VELOCITY IS DETERMINED

Velocity can be measured by using a speedometer in combination with some sort of device that indicates the instantaneous direction of travel. In a car, this can be a magnetic compass. In high-end vehicles, GPS receiving equipment can indicate the instantaneous direction in which motion occurs. But in a strict sense, even a speedometer and a compass or GPS receiver don't tell the whole story unless you're driving on a flat plain or prairie. In mid-state South Dakota, a speedometer and compass can define the instantaneous velocity of your car, but when you get into the Black Hills, you'll have to include a *clinometer* (a device for measuring the steepness of the grade you're ascending or descending) to get an absolutely perfect indication of your velocity.

Two-dimensional direction components can be denoted either as compass (azimuth) bearings, or as angles measured counterclockwise with respect to the axis pointing "east." The former system is preferred by hikers, navigators, and most people in real-world situations. The latter scheme is preferred by theoretical physicists and mathematicians. In Fig. 15-6, the azimuth bearings of vectors **a**, **b**, and **c** are approximately 90°, 130°, and 45°, respectively. These correspond to *points of the compass* called east (E), east-southeast (ESE), and northeast (NE).

A three-dimensional velocity vector consists of a magnitude component and two direction angles. One of the angles is the azimuth, and the other is called *elevation*, measured in degrees above the horizontal (positive angles) or below it (negative angles). The elevation angle can be as small as −90° (straight down) or as large as +90° (straight up). The horizontal direction is indicated by an elevation angle of 0°. Elevation angle should not be confused with elevation above or below sea level. That's an entirely different thing, measured in meters.

Acceleration

Acceleration is an expression of the rate of change in the velocity of an object. This can occur as a change in speed, a change in direction, or both. Acceleration can be defined in one dimension (along a straight line), in two dimensions (within a flat plane), or in three dimensions (in space), just as can velocity. Acceleration sometimes takes place in the same direction as an object's velocity vector, but this is not necessarily the case.

ACCELERATION IS A VECTOR

Acceleration, like velocity, is a vector quantity. Sometimes the magnitude of the acceleration vector is called "acceleration," and is usually symbolized by the lowercase italic letter a. But technically, the vector expression should be used; it is normally symbolized by the lowercase bold letter **a**.

In our previous example of a car driving along a highway, suppose the speed is constant at 25 m/s. The velocity changes when the car goes around

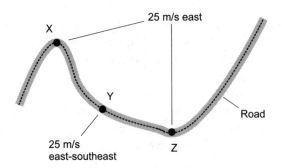

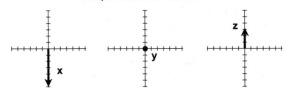

Fig. 15-7.   Acceleration vectors **x**, **y**, and **z** for a car at three points (X, Y, and Z) along a road. The magnitude of **y** is 0 because there is no acceleration at point Y.

curves, and also if the car crests a hilltop or bottoms-out in a ravine or valley (although these can't be shown in this two-dimensional drawing). If the car is going along a straight path, and its speed is increasing, then the acceleration vector points in the same direction that the car is traveling. If the car puts on the brakes, still moving along a straight path, then the acceleration vector points exactly opposite the direction of the car's motion.

Acceleration vectors can be graphically illustrated as arrows. Figure 15-7 illustrates acceleration vectors for a car traveling along a level, but curving, road at a constant speed of 25 m/s. Three points are shown, called X, Y, and Z. The corresponding acceleration vectors are **x**, **y**, and **z**. Because the speed is constant and the road is level, acceleration only takes place where the car encounters a bend in the road. At point Y, the road is essentially straight, so the acceleration is zero (**y** = 0). The zero vector is shown as a point at the origin of a vector graph.

HOW ACCELERATION IS DETERMINED

Acceleration magnitude is expressed in *meters per second per second*, also called *meters per second squared* (m/s^2). This seems esoteric at first. What does s^2 mean? Is it a "square second"? What in the world is that? Forget about trying to imagine it in all its abstract perfection. Instead, think of it

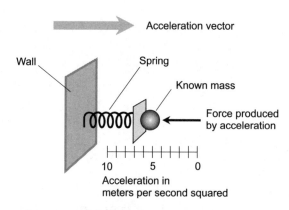

Fig. 15-8.   An accelerometer. This measures the magnitude only, and must be properly oriented to provide an accurate reading.

in terms of a concrete example. Suppose you have a car that can go from a standstill to a speed of 26.8 m/s in 5 seconds. Suppose that the acceleration rate is constant from the moment you first hit the gas pedal until you have attained a speed of 26.8 m/s on a level straightaway. Then you can calculate the acceleration magnitude:

$$a = (26.8 \text{ m/s})/(5 \text{ s}) = 5.36 \text{ m/s}^2$$

The expression s^2 translates, in this context, to "second, every second." The speed in the above example increases by 5.36 meters per second, every second.

Acceleration magnitude can be measured in terms of force against mass. This force, in turn, can be determined according to the amount of distortion in a spring. The force meter shown in Fig. 15-4 can be adapted to make an *acceleration meter*, more technically known as an *accelerometer*, for measuring acceleration magnitude.

Here's how a spring type accelerometer works. A functional diagram is shown in Fig. 15-8. Before the accelerometer can be used, it is calibrated in a lab. For the accelerometer to work, the direction of the acceleration vector must be in line with the spring axis, and the acceleration vector must point outward from the fixed anchor toward the mass. This produces a force on the mass. The force is a vector that points directly against the spring, exactly opposite the acceleration vector.

A common weight scale can be used to indirectly measure acceleration. When you stand on the scale, you compress a spring or balance a set of masses on a lever. This measures the downward force that the mass of your body exerts as a result of a phenomenon called the *acceleration of gravity*. The effect of gravitation on a mass is the same as that of an upward acceleration of approximately $9.8 \, \text{m/s}^2$. Force, mass, and acceleration are interrelated as follows:

$$\mathbf{F} = m\mathbf{a}$$

That is, force is the product of mass and acceleration. This formula is so important that it's worth remembering, even if you aren't a scientist. It quantifies and explains a lot of things in the real world, such as why it takes a fully loaded semi truck so much longer to get up to highway speed than the same truck when it's empty, or why, if you drive around a slippery curve too fast, you risk sliding off the road.

Suppose an object starts from a dead stop and accelerates at an average magnitude of a_{avg} in a straight line for a period of time t. Suppose after this length of time, the distance from the starting point is d. Then this formula applies:

$$d = a_{\text{avg}} t^2 / 2$$

In the above example, suppose the acceleration magnitude is constant; call it a. Let the instantaneous speed be called v_{inst} at time t. Then the instantaneous speed is related to the acceleration magnitude as follows:

$$v_{\text{inst}} = at$$

PROBLEM 15-9
Suppose two objects, denoted by curves A and B in Fig. 15-9, accelerate along straight-line paths. What is the instantaneous acceleration a_{inst} at $t = 4$ seconds for object A?

SOLUTION 15-9
The acceleration depicted by curve A is constant, because the speed increases at a constant rate with time. (That's why the graph is a straight line.) The number of meters per second squared does not change throughout the time span shown. In 10 seconds, the object accelerates from $0 \, \text{m/s}$ to $10 \, \text{m/s}$; that's a rate of speed increase of one meter per second per second ($1 \, \text{m/s}^2$). Therefore, the acceleration at $t = 4$ seconds is $a_{\text{inst}} = 1 \, \text{m/s}^2$.

PROBLEM 15-10
What is the average acceleration a_{avg} of the object denoted by curve A in Fig. 15-9, during the time span from $t = 2$ seconds to $t = 8$ seconds?

SOLUTION 15-10
Because the curve is a straight line, the acceleration is constant; we already know it is $1 \, \text{m/s}^2$. Therefore, $a_{\text{avg}} = 1 \, \text{m/s}^2$ between any two points in time shown in the graph, including the two points corresponding to $t = 2$ seconds and $t = 8$ seconds.

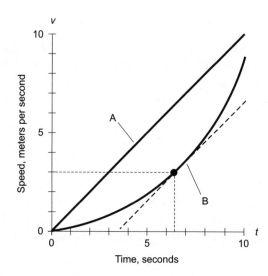

Fig. 15-9. Illustration for Problems 15-9 through 15-13.

PROBLEM 15-11

Examine curve B in Fig. 15-9. What can be said about the instantaneous acceleration of the object whose motion is described by this curve?

SOLUTION 15-11

The object starts out accelerating slowly, and as time passes, its instantaneous rate of acceleration increases.

PROBLEM 15-12

Use visual approximation in the graph of Fig. 15-9. At what time t is the instantaneous acceleration a_{inst} of the object described by curve B equal to $1 \, \text{m/s}^2$?

SOLUTION 15-12

Use a straight-edge to visualize a line tangent to curve B whose slope is the same as that of curve A. Then locate the point on curve B where the line touches curve B. Finally, draw a line straight down, parallel to the speed (v) axis, until it intersects the time (t) axis. Read the value off the t axis. Here, it appears to be about $t = 6.3$ seconds.

PROBLEM 15-13

Use visual approximation in the graph of Fig. 15-9. Consider the object whose motion is described by curve B. At the point in time t where the instantaneous acceleration a_{inst} is $1 \, \text{m/s}$, what is the instantaneous speed, v_{inst}, of the object?

SOLUTION 15-13
Locate the same point that you found in Problem 15-12, corresponding to the tangent point of curve B and the line parallel to curve A. Draw a horizontal line to the left, until it intersects the speed (v) axis. Read the value off the v axis. In this example, it looks like it's about $v_{inst} = 3.0\,\text{m/s}$.

Momentum

Momentum is an expression of "heft in motion," the product of an object's mass and its velocity. Momentum, like velocity, is a vector quantity, and its magnitude is expressed in kilogram meters per second (kg · m/s).

MOMENTUM AS A VECTOR

Suppose the speed of an object (in meters per second) is v, and the mass of the object (in kilograms) is m. Then the magnitude of the momentum, p, is their product:

$$p = mv$$

This is not the whole story. To fully describe momentum, the direction, as well as the magnitude, must be defined. That means we must consider the velocity of the mass in terms of its speed and direction. (A 2-kg brick flying through your east window is not the same as a 2-kg brick flying through your north window.) If we let **v** represent the velocity vector and **p** represent the momentum vector, then we can say this:

$$\mathbf{p} = m\mathbf{v}$$

IMPULSE

The momentum of a moving object can change in three different ways:

- A change in the mass of the object.
- A change in the speed of the object.
- A change in the direction of the object's motion.

Let's consider the second and third of these possibilities together; then this constitutes a change in the velocity.

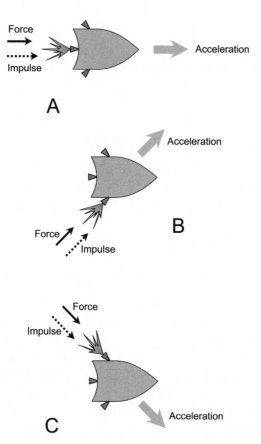

Fig. 15-10.   Three ways in which an impulse can cause a space ship to accelerate. At A, getting the ship to move straight ahead at higher speed. At B and C, getting the ship to turn.

Let's put our everyday-world time clock into "future mode." Imagine a massive space ship, coasting along a straight-line path in interstellar space. Consider a point of view, or *reference frame*, such that the velocity of the ship can be expressed as a vector pointing in a certain direction. A force **F** can be applied to the ship by firing a rocket engine. Imagine that there are several engines on the ship, one intended for driving it forward at increased speed, and others capable of changing the vessel's direction. If any engine is fired for t seconds with a force vector of **F** newtons (as shown by the three examples in Fig. 15-10), then the product **F**t is called the *impulse*. Impulse is a vector, symbolized by the uppercase boldface letter **I**, and its magnitude is expressed in kilogram meters per second (kg $\cdot$ m/s). Here's the formula:

$$\mathbf{I} = \mathbf{F}t$$

Impulse on an object always produces a change in the object's velocity. That's a good thing; it is the purpose of the rocket engines in our space ship! Recall the above formula concerning mass m, force $\mathbf{F}$, and acceleration $\mathbf{a}$:

$$\mathbf{F} = m\mathbf{a}$$

Substitute $m\mathbf{a}$ for $\mathbf{F}$ in the equation for impulse. Then we get this:

$$\mathbf{I} = (m\mathbf{a})t$$

Now remember that acceleration is a change in velocity per unit time. Suppose the velocity of the space ship is $\mathbf{v}_1$ before the rocket is fired, and $\mathbf{v}_2$ afterwards. Then, assuming the rocket engine produces a constant force while it is fired:

$$\mathbf{a} = (\mathbf{v}_2 - \mathbf{v}_1)/t$$

We can substitute in the previous equation to get

$$\mathbf{I} = m((\mathbf{v}_2 - \mathbf{v}_1)/t)t = m\mathbf{v}_2 - m\mathbf{v}_1$$

This means the impulse is equal to the change in momentum.

We have just derived an important law of real-world motion. Impulse is expressed in kilogram-meters per second (kg·m/s), just as is momentum. You might think of impulse as momentum in another form. When an object is subjected to an impulse, the object's momentum vector $\mathbf{p}$ changes. The vector $\mathbf{p}$ can grow larger or smaller in magnitude, or it can change direction, or both of these things can happen.

PROBLEM 15-14
Suppose a truck having a mass of 2000 kg moves at a constant speed of 20 m/s in a northerly direction. A braking impulse, acting in a southerly direction, slows the truck down to 15 m/s, but it still moves in a northerly direction. What is the impulse responsible for this change in speed?

SOLUTION 15-14
The original momentum, $\mathbf{p}_1$, is the product of the mass and the initial velocity:

$$\mathbf{p}_1 = 2000\,\text{kg} \times 20\,\text{m/s} = 40{,}000\,\text{kg} \cdot \text{m/s}$$

in a northerly direction

The final momentum, $\mathbf{p}_2$, is the product of the mass and the final velocity:

$$\mathbf{p}_2 = 2000\,\text{kg} \times 15\,\text{m/s} = 30{,}000\,\text{kg} \cdot \text{m/s}$$

in a northerly direction

Thus, the change in momentum is $\mathbf{p}_2 - \mathbf{p}_1$:

$$\mathbf{p}_2 - \mathbf{p}_1 = 30{,}000 \, \text{kg} \cdot \text{m/s} - 40{,}000 \, \text{kg} \cdot \text{m/s} = -10{,}000 \, \text{kg} \cdot \text{m/s}$$

in a northerly direction

That is the same as $10{,}000 \, \text{kg} \cdot \text{m/s}$ in a southerly direction. Because impulse is the same thing as the change in momentum, the impulse is $10{,}000 \, \text{kg} \cdot \text{m/s}$ in a southerly direction.

Don't let this result confuse you. A vector with a magnitude of $-x$ in a certain direction is, in the real world, a vector with magnitude $+x$ in the exact opposite direction. Problems sometimes work out to yield vectors with negative magnitude. When that happens, reverse the direction and change the sign of the magnitude from minus to plus.

Quiz

Refer to the text in this chapter if necessary. A good score is eight correct. Answers are in the back of the book.

1. If you drive a 1000-kg vehicle down the road at 30 m/s, what is the momentum?
    (a) $33 \, \text{kg} \cdot \text{m/s}$
    (b) $1030 \, \text{kg} \cdot \text{m/s}$
    (c) $3000 \, \text{kg} \cdot \text{m/s}$
    (d) $30{,}000 \, \text{kg} \cdot \text{m/s}$

2. Why is it unwise to drive a pickup truck full of loose bricks?
    (a) If you suddenly have to swerve, the bricks' momentum could cause some of them to suddenly fly over the side onto the road or into the ditch.
    (b) If you suddenly have to stop, the bricks' momentum could cause some of them to sail through the back window and hit you.
    (c) If you accelerate forward at too great a rate, the bricks' momentum could cause some of them to be dumped off the back of the truck and onto the road.
    (d) All of the above.

3. If you drive a truck full of loose bricks down a straight, level stretch of highway, the momentum vector of each brick points
    (a) in the same direction as the truck travels
    (b) in the opposite direction from that in which the truck travels

(c) at a right angle to the direction in which the truck travels

(d) nowhere, because momentum is not a vector

4. Imagine that a boat whose mass is 500 kg (with you and all your gear in it) is traveling straight forward on a lake at 10 m/s. You cut the engine and lift it out of the water, so you are coasting straight forward at 10 m/s. What impulse vector will bring the boat to a stop? Assume the boat moves along the water with no friction, and that there is no wind or current.

    (a) 50 kg·m/s, pointing forward.

    (b) 50 kg·m/s, pointing backward.

    (c) 5000 kg·m/s, pointing forward.

    (d) 5000 kg·m/s, pointing backward.

5. Suppose the town of Blissville is 5400 km from the town of Happyton, as measured along the Interstate highways. You drive from Blissville to Happyton over the course of several days, and you keep track of your total driving time. You come to the conclusion that you were traveling at an average speed of 15 m/s. How many hours were you driving?

    (a) 19

    (b) 100

    (c) 360

    (d) It is impossible to answer this without more information.

6. Imagine the space-ship scenario described in the text and illustrated in Fig. 15-10. Suppose a rocket engine is to be installed in a position and orientation such that, when it is fired, it causes the ship to slow down but continue in the same direction. Where should such a rocket be placed, and in what direction should its exhaust come out?

    (a) It should be placed on the nose of the ship, and its exhaust should come out directly forward.

    (b) It should be placed on the nose of the ship, and its exhaust should come out directly backward.

    (c) It should be placed on the back end of the ship, and its exhaust should come out directly sideways.

    (d) Any of the above.

7. Suppose you buy a car and the manufacturer claims that it can accelerate from 0 to 30 m/s in 6 seconds. Suppose you take the car for a test drive on a straight, level, dry stretch of roadway on a racetrack (with the permission of the track owner and under the supervision of the highway patrol, of course!). You discover that the manufacturer's

claim is true. What is the average acceleration rate during the 6-second period during which your car's speed has increased from a dead stop to 30 m/s?

(a) 5 m/s^2

(b) 30 m/s^2

(c) 180 m/s^2

(d) It is impossible to answer this without more information.

8. Suppose you have a mass of 60 kg on the earth's surface. The moon has a gravitational field that is 1/6 as strong, at the surface, as the earth's gravitational field. If you travel to the moon, what will be your mass there?

(a) 10 kg

(b) 60 kg

(c) 360 kg

(d) It is impossible to know without more information.

9. If you stand on the surface of the earth, your mass vector points

(a) straight down

(b) straight up

(c) horizontally

(d) nowhere, because mass is not a vector

10. Suppose you get a new set of tires for your truck, and they are larger in diameter than the tires you have been using. You know that your odometer was accurate with your old tires. If you drive 100 km with the new tires, your odometer, which measures distance according to the number of axle rotations, will tell you that you have traveled

(a) less than 100 km

(b) exactly 100 km

(c) more than 100 km

(d) too fast

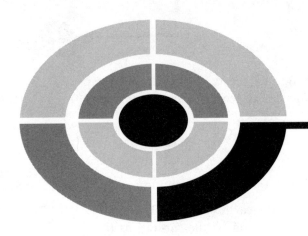

Test: Part 4

Do not refer to the text when taking this test. You may draw diagrams or use a calculator if necessary. A good score is at least 30 correct. Answers are in the back of the book. It's best to have a friend check your score the first time, so you won't memorize the answers if you want to take the test again.

1. Which of the following represents an arithmetic sequence?
    (a) 0, 1, 2, 4, 8, 16, 32
    (b) 2, 5, 8, 11, 14, 17, 20
    (c) 1, −1, 1, −1, 1, −1, 1
    (d) 1, 1/2, 1/3, 1/4, 1/5, 1/6, 1/7
    (e) 1, 1/2, 1/4, 1/8, 1/16, 1/32, 1/64

2. A car is advertised as being capable of going from 0 to 60 miles per hour in 5 seconds. What is the average acceleration, in miles per hour per second (mi/hr/s) that this figure represents?
    (a) 5 mi/hr/s
    (b) 10 mi/hr/s
    (c) 12 mi/hr/s

   (d) 300 mi/hr/s

   (e) More information is necessary to calculate it.

3. Suppose that when you "floor" the gas pedal in your car, it produces a certain constant forward force. This results in a certain forward acceleration when you're all alone in the car. If you fill up the car with passengers so its mass increases by 25% compared to its mass when you are alone in it, and then you "floor" the gas pedal, what happens to the forward acceleration? Ignore any possible effects of friction. Remember that force is the product of mass and acceleration.

   (a) The acceleration becomes 1.25 times as great as it is with only you in the car.

   (b) The acceleration stays the same as it is with only you in the car.

   (c) The acceleration becomes 4/5 times as great as it is with only you in the car.

   (d) The acceleration becomes 3/4 as great as it is with only you in the car.

   (e) The acceleration becomes half as great as it is with only you in the car.

4. Suppose that when you "floor" the gas pedal in your car, it produces a certain constant forward force. This results in a certain forward acceleration when you're all alone in the car. If you put high-performance fuel in the tank, producing a 25% increase in the force produced when you "floor" the gas pedal, what happens to the forward acceleration when you are alone in the car and you "floor" the gas pedal? Ignore any possible effects of friction. Remember that force is the product of mass and acceleration.

   (a) The acceleration stays the same as it is with the ordinary fuel.

   (b) The acceleration becomes 1.25 times as great as it is with the ordinary fuel.

   (c) The acceleration becomes 1.563 times as great as it is with the ordinary fuel.

   (d) The acceleration becomes twice as great as it is with the ordinary fuel.

   (e) The acceleration becomes four times as great as it is with the ordinary fuel.

5. A quantity that has magnitude and direction is known as

   (a) a variable

   (b) a constant

(c) a scalar

(d) a vector

(e) a logarithm

6. Imagine that you've purchased a new hi-fi audio amplifier for your car. It contains volume (or gain) control knobs, one for each channel, calibrated in numbers from 0 to 10. The specification sheet in the instruction manual for your new amplifier contains a graph that looks like Fig. Test 4-1. From this graph, it is apparent that

    (a) the output volume, in watts, is directly proportional to the volume control setting in either channel

    (b) the output volume, in decibels, is directly proportional to the volume control setting in either channel

    (c) the output volume, in watts, is inversely proportional to the volume control setting in either channel

    (d) the output volume, in decibels, is inversely proportional to the volume control setting in either channel

    (e) the output volume increases according to the square of the volume control setting in either channel

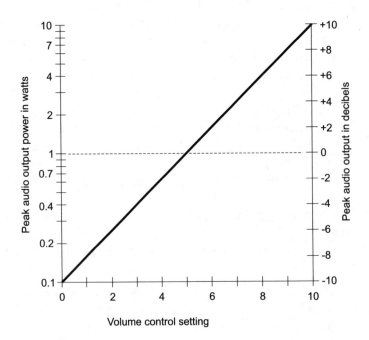

Fig. Test 4-1.   Illustration for Part Four Test Questions 6 through 8.

7. Suppose you're listening to music using the amplifier whose specifications include a graph such as that shown in Fig. Test 4-1. If you increase the volume control setting by 2 units, the peak output power, in watts, increases by
   (a) 0.2 watts
   (b) 0.4 watts
   (c) 2 watts
   (d) 4 watts
   (e) an amount that depends on the initial setting.

8. Suppose you're listening to music using the amplifier whose specifications include a graph such as that shown in Fig. Test 4-1. If you increase the volume control setting by 2 units, the peak output power, in decibels, increases by
   (a) 0.2 decibels
   (b) 0.4 decibels
   (c) 2 decibels
   (d) 4 decibels
   (e) an amount that depends on the initial setting

9. The sum of all the terms in a sequence is called
   (a) a partial sum
   (b) a series
   (c) an arithmetic sequence
   (d) a geometric sequence
   (e) a linear sequence

10. Which of the following is always a scalar quantity?
    (a) The velocity of a train.
    (b) The acceleration of a car.
    (c) The impulse produced by a rocket engine.
    (d) The mass of a truckload of bricks.
    (e) The force produced by the wind against a wall.

11. Suppose you're driving a motorboat across a river. The bow of the boat is pointing straight south, and your water speed is 7 knots. There is a steady current in the water moving from east to west at a speed of 7 knots. What is your direction (heading) relative to land?
    (a) South.
    (b) Southeast.
    (c) Southwest.
    (d) East.
    (e) West.

12. Suppose you're driving a motorboat across a river. The bow of the boat is pointing straight north, and your water speed is 12 knots. There is a steady current in the water moving from west to east at a speed of 9 knots. What is your speed relative to land?
(a) 3 knots.
(b) 9 knots.
(c) 12 knots.
(d) 15 knots.
(e) It is impossible to tell without more information.

13. Suppose you buy a house for $200,000. You are confident that its value will increase by exactly 10% per year for the next 10 years. According to this formula, what should the house be worth after 3 years?
(a) $230,000
(b) $242,000
(c) $260,000
(d) $266,200
(e) More information is necessary to answer this question.

14. Suppose you buy a house for $200,000, and its value falls by exactly 10% per year for 3 years. What is the house worth at the end of the 3 years?
(a) $170,000
(b) $162,000
(c) $145,800
(d) $140,000
(e) More information is necessary to answer this question.

15. Which of the following can be expressed as a vector quantity?
(a) The mass of a bag of apples.
(b) The brightness of a source of light.
(c) The loudness of a sound.
(d) The acceleration of a car.
(e) The volume of water in a swimming pool.

16. In Fig. Test 4-2, which of the vectors in the lower diagrams best represents the velocity vector for the car shown in the top diagram? Imagine the car moving generally from left to right along the road as shown. Assume each division represents 5 meters per second.
(a) **a**
(b) **b**
(c) **c**

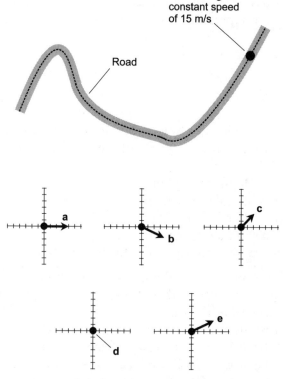

Fig. Test 4-2. Illustration for Part Four Test Questions 16 through 18.

(d) **d**

(e) **e**

17. In Fig. Test 4-2, which of the vectors in the lower diagrams best represents the acceleration vector for the car shown in the top diagram? Imagine the car moving generally from left to right along the road as shown. Assume each division represents 5 meters per second squared.

(a) **a**

(b) **b**

(c) **c**

(d) **d**

(e) **e**

18. Imagine the car reversing direction as shown in the diagram of Fig. Test 4-2, so it moves generally from right to left at a constant speed of 15 meters per second. Which of the vectors in the lower

diagrams best represents the acceleration vector in this case? Assume each division represents 5 meters per second squared.

(a) **a**

(b) **b**

(c) **c**

(d) **d**

(e) **e**

19. Suppose Jimsville is 60 miles from Joesville. You need to get from one town to the other in 90 minutes or less. Your minimum average speed

(a) must be 30 miles per hour

(b) must be 40 miles per hour

(c) must be 45 miles per hour

(d) must be 60 miles per hour

(e) cannot be determined from this information

20. Suppose Jimsburg is 80 kilometers from Joesburg. You make a trip from one town to the other in exactly 80 minutes. Your instantaneous acceleration

(a) must be at least 1 meter per second squared

(b) must be at least 60 meters per second squared

(c) must be at least 16.17 millimeters per second squared

(d) must be at least 16.67 meters per second squared

(e) cannot be determined from this information

21. Suppose you deposit exactly $65,536.00 into an account on the first day of the month. On the second day of the month, you deposit exactly half this amount. On the third day, you deposit half the amount that you deposited on the second day. You repeat this process until you can no longer cut the deposits in half while still making them in whole cents (that is, without "breaking up a penny"). What will be the amount you have in the bank at that time?

(a) $131,071.75

(b) $131,071.00

(c) $131,072.75

(d) $131,072.00

(e) That time will never be reached; the account balance will continue to increase without limit.

22. Speed can be expressed in terms of

(a) distance multiplied by time

(b) distance per unit time

(c) time per unit distance

(d) acceleration per unit time

(e) velocity multiplied by time

23. Any radioactive material has a specific period of time known as its *half-life*, over which its radioactivity drops by 50%. Suppose the half-life of element X is 100 years. Right now the radiation intensity from a sample of element X is at a certain level. In 100 years, the element X will be 50% as radioactive as it is right now. After another 100 years, the radioactivity of the element X will drop by another 50%. This process will repeat, century after century, eon after eon. The radiation-intensity levels of any given sample of element X, measured in constant "radiation units (RU)" at intervals of 100 years:

(a) increase without limit

(b) alternate between positive and negative

(c) comprise a geometric progression

(d) add up to zero

(e) converge toward 1

24. Suppose you want to check the floor in a room to see if it is perfectly horizontal. You know the floor is perfectly flat throughout the room. You have a carpenter's level with which you can check to see if any particular line along the floor is horizontal. What is the minimum number of readings you must make with the level in order to know whether or not the floor is perfectly horizontal?

(a) 1

(b) 2

(c) 3

(d) 4

(e) 5

25. Which of the following can be a unit of acceleration magnitude?

(a) Meters per second.

(b) Seconds per meter.

(c) Meters per second squared.

(d) Seconds per meter squared.

(e) Degrees per meter squared.

26. Imagine a room whose floor and ceiling are perfectly square, and whose four walls are all perfect rectangles. Imagine one corner of this room, where two walls intersect the floor. There are three line segments that come together at this corner: the line segment where

the two walls intersect, the line segment where one wall intersects the floor, and the line segment where the other wall intersects the floor. At the corner where they all come together, these three line segments are
(a) mutually parallel
(b) mutually perpendicular
(c) mutually skewed
(d) mutually equilateral
(e) none of the above

27. A flat geometric plane in 3D space can be uniquely defined by
(a) a point in the plane and a vector perpendicular to the plane
(b) three points that do not all lie on the same straight line
(c) two parallel straight lines
(d) two intersecting straight lines
(e) any of the above

28. Suppose you have a jar in which you keep change that accumulates in the course of your shopping adventures. Every day, you put 76 cents into the jar. The amount of money in the jar increases day by day, according to
(a) an arithmetic progression
(b) a geometric progression
(c) a logarithmic progression
(d) an exponential progression
(e) a convergent progression

29. Any radioactive material has a specific period of time known as its *half-life*, over which its radioactivity drops by 50%. Suppose the half-life of element *Y* is 80,000 years, someone has just handed you a sample of it, and the radiation level from it is 128 "radiation units (RU)." In 80,000 years, the sample will be half as radioactive as it is today. After another 80,000 years, the radioactivity will go down by half, again. This decay process will go on essentially forever. Suppose the maximum safe level for human exposure to any radioactive substance is 16 RU. How long will it be before it's no longer dangerous to hang around the piece of element *Y* you've just been given?
(a) 240,000 years.
(b) 160,000 years.
(c) 80,000 years.
(d) 40,000 years.
(e) 20,000 years.

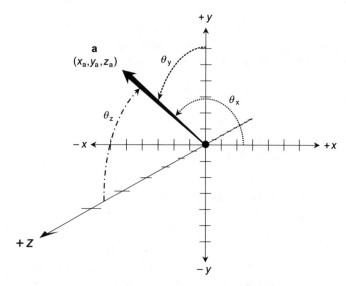

Fig. Test 4-3. Illustration for Part Four Test Questions 30 through 34.

30. What would Fig. Test 4-3 most likely represent?
    (a) The frequency setting of a hi-fi stereo radio receiver.
    (b) The direction in which a car or truck drives down a level street.
    (c) The direction in which a broadband satellite Internet dish is pointed.
    (d) The compass bearing of a hiker in the wilderness.
    (e) The acceleration of a speedboat on a lake.

31. In Fig. Test 4-3, what do θ_x, θ_y, and θ_z represent?
    (a) Direction cosines.
    (b) Direction angles.
    (c) Compass bearings.
    (d) Elevation angles.
    (e) Azimuth bearings.

32. In Fig. Test 4-3, what is represented by the lowercase, boldface letter **a**?
    (a) A ray of light.
    (b) An exponent.
    (c) A logarithm.
    (d) A scalar.
    (e) A vector.

33. What is another thing that Fig. Test 4-3 might depict?
    (a) The velocity at which a missile is fired.
    (b) Sunlight shining down on the surface of the earth.

(c) The motion of a train along a level track.
(d) The speed of the wind in a hurricane.
(e) The time and temperature in a certain place.

34. The diagram in Fig. Test 4-3 portrays?
(a) 1 spatial dimension.
(b) 2 spatial dimensions.
(c) 3 spatial dimensions.
(d) 4 spatial dimensions.
(e) 5 spatial dimensions.

35. The human ear and brain perceive the loudness of a sound
(a) in proportion to the square of the sound power
(b) in proportion to the base-10 exponential of the sound power
(c) in proportion to the base-e exponential of the sound power
(d) in proportion to the logarithm of the sound power
(e) in inverse proportion to the sound power

36. Suppose we are told the following sequence S is an arithmetic sequence:

$$S = 80, 75, x, y, z, 55, 50$$

What is the difference between x and z?
(a) 5
(b) 10
(c) 15
(d) 20
(e) It cannot be determined without more information.

37. Suppose you are told that the intensity of some effect Y varies according to the 3/2 power of the strength of some other effect X. If the strength of effect X doubles, by what factor does the intensity of effect Y increase?
(a) It does not change.
(b) It doubles.
(c) It quadruples.
(d) It increases by a factor of the square root of 8.
(e) More information is needed to answer this question.

38. Suppose you are building a coffee table. Your friend wants it to have four legs attached to the table at the corners of a square. But you want it to have only three legs, attached to the table at the vertices of an equilateral triangle. You know that a three-legged table

won't wobble, even if the floor is irregular or if the legs don't all turn out to be exactly the same length. On what geometric principle is this fact based?

(a) A vector normal to a flat plane cannot lie in that plane.
(b) Two intersecting, straight lines uniquely define a flat plane.
(c) Three points, not all on the same line, uniquely define a flat plane.
(d) Four points can lie in the same plane, but this is not always true.
(e) Four points can lie at the corners of a square, but this is not always true.

39. Suppose your computer receives (unknown to you) an e-mail worm that causes it to automatically send the same message to 100 other computers. Ninety percent (90%) of the computers that receive this e-mail block it out, but the other 10% get it and execute its nefarious instructions. This process continues indefinitely, with one iteration taking place every minute. The number of these useless, unethical, illegal e-mail messages clogging up the Internet increases, per unit minute, according to

(a) an arithmetic progression
(b) a logarithmic progression
(c) a geometric progression
(d) a quadratic progression
(e) a cubic progression

40. Suppose you step on the "gas pedal" of a high-performance car while you take a banked curve on a race track. The magnitude and direction of the acceleration of the car at a particular instant in time can be fully defined as

(a) a vector
(b) a logarithm
(c) a constant
(d) a trigonometric function
(e) an exponent

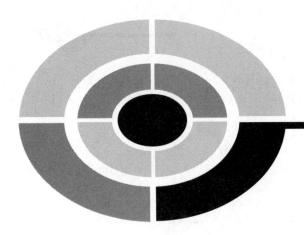

Final Exam

Do not refer to the text when taking this test. You may draw diagrams or use a calculator if necessary. A good score is at least 75 correct. Answers are in the back of the book. It's best to have a friend check your score the first time, so you won't memorize the answers if you want to take the test again.

1. Which of the following statements is false?
   (a) A dependent variable can change in value, but its value is affected by at least one other factor.
   (b) An independent variable can change in value, but its value is not influenced by anything else in a given scenario.
   (c) A mathematical function is always a mathematical relation.
   (d) A mathematical relation is always a mathematical function.
   (e) A function always has at least one dependent variable.

2. What is the product of 2/3 and 4/5?
   (a) 6/15
   (b) 8/15
   (c) 3/4

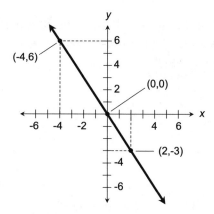

Fig. Exam-1. Illustration for Final Exam Question 3.

(d) 1
(e) 22/15

3. What is the equation shown by the straight line in Fig. Exam-1?
   (a) $3x + 2y = 0$
   (b) $2x + 3y = 0$
   (c) $-2x - 3y = 0$
   (d) $-4x - 6y = 0$
   (e) $6x + 3y = 0$

4. Consider the following equation:

$$(y^2 - 2y + 1)(y + 4)(y + 5)(y + 6) = 0$$

   How many real-number solutions are there to this equation?
   (a) 1
   (b) 2
   (c) 3
   (d) 4
   (e) 5

5. It is known that 1 foot is equal to approximately 30.5 centimeters, and that 1 inch is exactly 1/12 of a foot. Based on these facts, how many centimeters are there in 1 inch?
   (a) 0.305
   (b) 0.393
   (c) 2.54
   (d) 3.28
   (e) More information is needed to answer this question.

6. Suppose you flip an unbiased coin, and it comes up "heads" 4 times in a row. If you flip it a fifth time, what is the probability that it will come up "heads" again?
   (a) 50%
   (b) 25%
   (c) 12.5%
   (d) 6.25%
   (e) 3.125%

7. A nomograph
   (a) consists of two graduated axes that intersect at a right angle
   (b) consists of three graduated axes that intersect at right angles
   (c) consists of two graduated scales lined up directly with each other
   (d) cannot be used to compare variables such as temperatures in different scales
   (e) consists of points in a coordinate plane, connected by straight lines

8. If the rectangular-coordinate graph of an equation is a parabola opening downward, the equation is
   (a) linear
   (b) circular
   (c) quadratic
   (d) cubic
   (e) quartic

9. Dionysus is thinking of a real number. If he adds 7 to this number, he gets the same result as if the number is doubled. What is the number?
   (a) 7
   (b) 2
   (c) 7/2
   (d) 2/7
   (e) There is no such real number.

10. Imagine the set W of days in a week. Tuesday is
    (a) a subset of W
    (b) a proper subset of W
    (c) an element of W
    (d) a fraction of W
    (e) a point of W

11. Which of the following (a), (b), (c), or (d) is not a real number?
    (a) $4^{1/2}$
    (b) $4^{1/3}$
    (c) $4^{1/4}$

(d) $4^{-1/5}$

(e) All of the above are real numbers.

12. In probability, the set of all possible outcomes in the course of an experiment is called
    (a) the dependent variable
    (b) the independent variable
    (c) the sample space
    (d) the population
    (e) the event

13. Suppose your history teacher tells you that there is no good evidence that the ancient Roman emperor Nero played a musical instrument while the town burned. The teacher says, "It is nothing but a wild tale. The probability is only 10% that Nero actually played a musical instrument while Rome was on fire." This last statement is an example of
    (a) a sample space that is too small
    (b) conflicting data
    (c) the "probability fallacy"
    (d) mistaking a dependent variable for an independent variable
    (e) mistaking a permutation for a combination

14. Imagine a quantity x. Suppose you add 7 to x, and get the same result as if you subtract 5 from x. What is the real-number value of x?
    (a) 7
    (b) 5
    (c) 7/5
    (d) 5/7
    (e) The quantity x is not a real number.

15. Suppose two corporations want to advertise on a billboard. The billboard has a trapezoidal shape, as shown in Fig. Exam-2. The corporations decide that the most equitable way to divide up the space is to divide it horizontally across the middle, so the top portion (light shading) and bottom portion (dark shading) are of equal height. The dimensions are as shown. Assume they are exact to the nearest millimeter. How long is the central, horizontal line that divides the light and dark shaded regions?
    (a) More information is necessary to answer this.
    (b) 11.750 meters.
    (c) 11.833 meters.

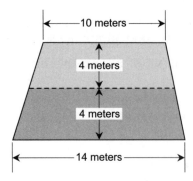

Fig. Exam-2. Illustration for Final Exam Questions 15 and 16.

    (d) 12.000 meters.
    (e) 12.250 meters.

16. In the situation shown by Fig. Exam-2, the area of the top portion (light shading) of the billboard represents
    (a) 28.57% of the total billboard area
    (b) 33.33% of the total billboard area
    (c) 40.00% of the total billboard area
    (d) 45.83% of the total billboard area
    (e) less than half of the total billboard area, but more information is necessary to calculate the exact percentage

17. Suppose you have a standard 52-card deck. The deck is complete, and there is exactly one of each card (4 suits with 13 cards in each suit, and no jokers). You draw a card from this deck, then put the card back, shuffle the deck several times, and then draw another card. What is the probability that you'll get the same card twice?
    (a) 1/140,608
    (b) 1/2704
    (c) 1/52
    (d) 1/13
    (e) 1/4

18. What is the value of $3 + 5 \times 1^{-5} - 7$?
    (a) The expression is ambiguous without parentheses telling us in what order we should perform the operations.
    (b) -56
    (c) 56
    (d) -3
    (e) 1

19. Suppose a straight, flat, two-lane highway crosses a straight set of railroad tracks, such that the acute angle between the rails and the center line of the highway measures 60°. What is the measure of the obtuse angle between the rails and the center line of the highway?
    (a) 150°
    (b) 120°
    (c) 90°
    (d) 60°
    (e) 30°

20. Nikolai is thinking of a positive whole number. If the number is squared and then 10 is subtracted from the result, the difference is equal to 6. What is the number?
    (a) 2
    (b) 4
    (c) 6
    (d) 10
    (e) 100

21. Imagine the set F of the days in February. The set of all Sundays in February is
    (a) a subset of F
    (b) a superset of F
    (c) an element of F
    (d) a fraction of F
    (e) a point of F

22. Suppose your house has 2700 square feet of floor space, and the ceilings are all exactly 8 feet high. How many cubic feet are in your house?
    (a) 2700
    (b) 5400
    (c) 10,800
    (d) 21,600
    (e) More information is needed to answer this question.

23. Suppose your house has 2700 square feet of floor space, and the ceilings are all exactly 8 feet high. How many cubic yards are in your house? (A linear yard is exactly 3 linear feet.)
    (a) 800
    (b) 2400
    (c) 7200

(d) 14,400

(e) More information is needed to answer this question.

24. In statistics, a variable that can only attain certain specific values is called
    (a) a constant variable
    (b) an independent variable
    (c) a dependent variable
    (d) a discrete variable
    (e) a continuous variable

25. Not too long ago, a computer was considered fast if its processor worked at 1 MHz (1,000,000 cycles per second). A few years later, processor speeds had increased to 1 GHz (1,000,000,000 cycles per second). This represented a processor-speed increase of
    (a) 1 order of magnitude
    (b) 2 orders of magnitude
    (c) 3 orders of magnitude
    (d) 6 orders of magnitude
    (e) 9 orders of magnitude

26. A text discussion talks about an angle θ, and then asks you to assume that $\theta = \pi/2$. What is the equivalent measure of θ in angular degrees?
    (a) $45°$
    (b) $90°$
    (c) $135°$
    (d) $180°$
    (e) $270°$

27. Fig. Exam-3 shows points plotted in
    (a) polar coordinates
    (b) rectangular coordinates
    (c) conical coordinates
    (d) navigator's coordinates
    (e) Einstein coordinates

28. What are the coordinates of point R in Fig. Exam-3?
    (a) $(3, -4)$
    (b) $(3, -5)$
    (c) $(-3, 4)$
    (d) $(-3, 5)$
    (e) $(-4, 5)$

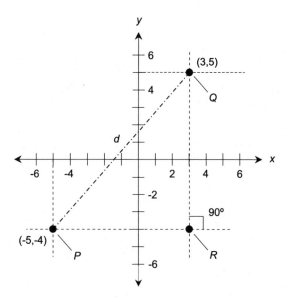

Fig. Exam-3. Illustration for Final Exam Questions 27 through 29.

29. What is the slope of the line connecting points Q and R (not points Q and P!) in Fig. Exam-3?
    (a) 0
    (b) 8/9
    (c) 9/8
    (d) It is not defined.
    (e) More information is needed to answer this question.

30. The year 2003 contained precisely 365 solar days. One solar day is exactly 24 hours long. How many seconds were there in the year 2003?
    (a) 5.256×10^5
    (b) 5.256×10^{-5}
    (c) 3.1536×10^7
    (d) 3.1536×10^{-7}
    (e) More information is needed to answer this question.

31. In the expression $w + xy^2$, which operation should be performed last when the values of w, x, and y are known?
    (a) The addition.
    (b) The multiplication.
    (c) The squaring (raising to the power of 2).

(d) It doesn't matter.

(e) There is no way to know, unless parentheses are included that tell us the order in which we should do the operations.

32. The mean for a discrete variable in a statistical distribution is
   (a) the smallest value
   (b) the largest value
   (c) the mathematical average of all the values
   (d) the value in the exact middle
   (e) the value that occurs most often

33. Suppose you are exactly 6 feet tall. Your doctor tells you not to go out in the sun unless your shadow is at least 10 feet long on level ground. This means the sun must be below a certain angle relative to the horizon. What is this angle? Round the answer off to the nearest angular degree.
   (a) $31°$
   (b) $37°$
   (c) $53°$
   (d) $59°$
   (e) More information is needed to answer this question.

34. Hypatia loves green apples. Every time her big sister, Katyusha, goes to the grocery store, Hypatia reminds her to get plenty of green apples. One day Katyusha comes back from the store and says, "Today they would sell me only a certain amount of green apples. They said if I weighed them, the kilograms and pounds scales had to say the same thing." Hypatia, sensing the truth immediately, says:
   (a) "How will we ever be able to eat all those apples before they spoil?"
   (b) "Oh, so they didn't have any green apples this week!"
   (c) "That's not saying anything. The pounds and kilograms scales always agree, no matter what you put on them."
   (d) "That will be enough so we can each have one apple a day for the next week."
   (e) "That's ridiculous. You can't use a kilograms scale to weigh anything."

35. Imagine a city in which streets run east–west, and avenues run north–south. Each street is numbered north and south from Center Street, and each avenue is numbered east and west from Broadway. Consecutively numbered streets and avenues are exactly 100 meters apart, so each city block measures 100 meters square. There are no

diagonal routes, and no routes are blocked off. You are at the intersection of 8th Street North and 6th Avenue East. What is the minimum distance you must drive, along the streets and avenues, to get to the intersection of 4th Street South and 3rd Avenue West?
(a) 1.1 kilometers.
(b) 1.2 kilometers.
(c) 1.5 kilometers.
(d) 2.0 kilometers.
(e) 2.1 kilometers.

36. In the above-mentioned city (Question 35), what is the straight-line distance that an unimpeded ray of light would travel to get from the intersection of 8th Street North and 6th Avenue East to the intersection of 4th Street South and 3rd Avenue West?
(a) 1.1 kilometers.
(b) 1.2 kilometers.
(c) 1.5 kilometers.
(d) 2.0 kilometers.
(e) 2.1 kilometers.

37. There is a well-known formula for converting degrees Fahrenheit to degrees Celsius. The formula is as follows:

$$C = (5/9)(F - 32)$$

where C represents the temperature in degrees Celsius (°C), and F represents the temperature in degrees Fahrenheit (°F). Normal human body temperature is considered to be 98.6°F. What is this on the Celsius scale?
(a) 22.8°C
(b) 37.0°C
(c) 209°C
(d) 235°C
(e) More information is needed to answer this question.

38. Suppose the economy of a country undergoes slow but steady inflation, at the rate of 2% per year. If an item costs exactly $1.00 in US dollars (or 100¢ in US cents) today in that country, how much will it cost in 10 years, assuming the rate of inflation remains constant and the price of that item rises in accordance with the rate of inflation? Round the answer off to the nearest cent.
(a) $1.10
(b) $1.22

(c) 90¢

(d) 82¢

(e) More information is needed to answer this question.

39. Suppose the economy of a country undergoes slow but steady infla-tion, at the rate of 2% per year. If $1.00 in US dollars (that is, 100 US cents) will buy 1 kilogram of cranberries today in that country, how many kilograms of cranberries will $1.00 buy in 10 years, assum-ing the rate of inflation remains constant and the price of cranberries rises in accordance with the rate of inflation? Round the answer off to the nearest 10 grams.

(a) 1.10 kilograms.

(b) 1.22 kilograms.

(c) 0.90 kilograms.

(d) 0.82 kilograms.

(e) More information is needed to answer this question.

40. Which of the following unit pairs can most reasonably represent the variables in a polar coordinate system?

(a) Radians and angular degrees.

(b) Angular degrees and meters.

(c) Meters and centimeters.

(d) Kilometers and feet.

(e) Sines and cosines.

41. You're watching a major-league baseball game on television. At the top of the screen is a line of data numbers, including the score, the ball-strike count, and the speed of every pitch in miles per hour. You know there are 5280 feet in a mile and 3600 seconds in an hour. How many feet per second is a 90-mile-per-hour pitch?

(a) More information is needed to answer this question.

(b) 56 feet per second.

(c) 61 feet per second.

(d) 132 feet per second.

(e) 145 feet per second.

42. If you drive a 2000-kg vehicle down the road at 20 m/s, what is the momentum?

(a) 40 kg · m/s

(b) 100 kg · m/s

(c) 2020 kg · m/s

(d) 40,000 kg · m/s

(e) More information is needed to answer this question.

43. Suppose you have a large sheet of paper that measures exactly 1 meter square, so its surface area is 1 square meter on either side. If you fold this piece of paper in half twice, what is the surface area, in millimeters, of one face of the resulting 4-ply folded sheet?
    (a) 1,000,000 square millimeters.
    (b) 250,000 square millimeters.
    (c) 125,000 square millimeters.
    (d) 62,500 square millimeters.
    (e) 31,250 square millimeters.

44. Suppose you are building a stairway from the ground floor to the second floor hallway in a house. The base of the stairwell is to be 5.245 meters from the base of the door at the top of the stairway. The base of this door is 3.105 meters above the level of the first floor. At what angle, relative to the vertical wall (not the floor), will the banisters on the stairway slant? Assume the banisters are exactly parallel to the line representing the slant of the stairwell. Assume the floors are all flat and level. Round the answer off to the nearest tenth of a degree.
    (a) 30.6°
    (b) 36.3°
    (c) 53.7°
    (d) 59.4°
    (e) More information is needed to answer this question.

45. Suppose you are building a stairway from the back yard to the second floor living room in a house. The base of the stairwell is to be 5.245 meters from the base of the door at the top of the stairway. The base of this door is 3.105 meters above the level of the ground. At what angle, relative to the vertical wall (not the floor), will the banisters on the stairway slant? Assume the banisters are exactly parallel to the line representing the slant of the stairwell. Assume that the ground is not flat and level near the house, but instead, slopes downhill away from the house. Round the answer off to the nearest tenth of a degree.
    (a) 30.6°
    (b) 36.3°
    (c) 53.7°
    (d) 59.4°
    (e) More information is needed to answer this question.

46. In a vertical bar graph, the value of a function's dependent variable, for a given value of the independent variable, is proportional to
(a) the height of a rectangle or strip
(b) the sharpness of a curve
(c) the width of a straight line
(d) the area of a pizza-slice-shaped wedge
(e) the area of a triangle

47. Let P_1 be the visible-light power (in watts) emitted by lamp number 1. Let P_2 be the visible-light power (also in watts) emitted by lamp number 2. Their perceived intensity relationship R, in decibels (dB), is given by this formula:

$$R = 10 \ \log \left(P_2/P_1\right)$$

Suppose lamp no. 1 emits 10 watts of visible light, while lamp no. 2 emits 40 watts of visible light. How many decibels brighter is lamp no. 2 than lamp no. 1?
(a) 3 dB
(b) 4 dB
(c) 6 dB
(d) 30 dB
(e) More information is needed to answer this question.

48. Suppose a car is traveling northwest at 15 meters per second. If this is expressed as a vector quantity, then "15 meters per second" describes the
(a) orientation of the vector
(b) velocity of the vector
(c) acceleration of the vector
(d) magnitude of the vector
(e) scalar of the vector

49. The set of points equidistant from the origin of a polar coordinate system is
(a) a straight line
(b) a parabola
(c) a hyperbola
(d) a circle
(e) a square

50. The set of points equidistant from the origin of a rectangular coordinate system is
(a) a straight line

(b) a parabola
(c) a hyperbola
(d) a circle
(e) a square

51. The mode for a discrete variable in a statistical distribution is
    (a) the smallest value
    (b) the largest value
    (c) the mathematical average of all the values
    (d) the value in the exact middle
    (e) the value that occurs most often

52. Suppose you are driving a car north on a back road at 15 meters per
    second. You see a squirrel in the road and, because there is no one
    driving behind you, you slam on the brakes. In exactly 3 seconds,
    you come to a complete stop, and the squirrel, paralyzed with fear
    in the middle of the road, is spared. Your change in speed can be con-
    sidered acceleration in a southerly direction (that is, contrary to the
    direction in which you are driving), at an average rate of
    (a) 45 meters per second per second
    (b) 18 meters per second per second
    (c) 12 meters per second per second
    (d) 5 meters per second per second
    (e) 3 meters per second per second

53. Which of the following functions looks like a sinusoid when graphed?
    (a) Cosine.
    (b) Cosecant.
    (c) Tangent.
    (d) Cotangent.
    (e) Secant.

54. Which of the following (a), (b), (c), or (d) is a legitimate way to
    manipulate an equation in which the variable is z?
    (a) Subtract 1 from the quantities on both sides of the equals sign.
    (b) Divide the quantities on both sides of the equals sign by 2.
    (c) Multiply the quantities on both sides of the equals sign by 10.
    (d) Add z to the quantities on both sides of the equals sign.
    (e) All of the above are legitimate ways to manipulate an equation in
        which the variable is z.

55. The square root of a negative integer is
    (a) a rational number
    (b) a real number

(c) an irrational number

(d) an imaginary number

(e) an undefined quantity

56. Suppose Happyton and Blissville are 240 miles apart on Route X-99. What is the average speed you must drive if you want to get from Happyton to Blissville in 8 hours?

(a) 30 meters per second.

(b) 30 feet per second

(c) 30 miles per hour

(d) 30 kilometers per hour.

(e) None of the above.

57. Suppose you pay a certain amount per month for your medical insurance right now. If the premiums increase by exactly 15% annually for the next 20 years, how many times as large will your monthly premium be in 20 years, as compared with what it is today?

(a) 3 times as large.

(b) 4 times as large.

(c) 11.5 times as large.

(d) 16.4 times as large.

(e) 23 times as large.

58. Suppose you want to paint the floor with two coats in a room that measures exactly 10 by 10 feet. One liquid quart of paint is advertised to cover 55 square feet. There are four liquid quarts in a liquid gallon. How much paint should you buy, in order to ensure that you can completely cover the floor with two coats, but while also ensuring that you waste the smallest possible amount of paint?

(a) 1 quart.

(b) 2 quarts.

(c) 3 quarts.

(d) 1 gallon.

(e) 1 gallon and 1 quart.

59. The force F (in newtons) required to accelerate an object of mass m (in kilograms) in a straight line at a rate of a (in meters per second squared) is given by the well-known equation:

$$F = ma$$

Imagine you are driving a truck whose mass is 5 metric tons. (A metric ton is 1000 kilograms.) The truck is accelerating forward on a level, flat, straight road at exactly 200 centimeters per second per second. What is the forward force causing this acceleration? Neglect the effects of friction, and assume there is no wind.

(a) 1000 newtons.

(b) 10,000 newtons.

(c) 100,000 newtons.

(d) 1,000,000 newtons.

(e) More information is needed to answer this question.

60. Suppose you graph two different equations in two variables. Both of the equations have graphs that look like parallel, straight lines. How many common solutions are there to this pair of equations?

(a) There are no solutions.

(b) There is one solution.

(c) There is more than one solution.

(d) There are infinitely many solutions.

(e) More information is needed to answer this.

61. Which of the following numbers is expressed in binary form?

(a) 2E5F

(b) 722

(c) 989880

(d) 10101

(e) 123210

62. Suppose you're sailing a motorboat from south to north across a river that runs exactly east–west. If there were no current in the river, your boat would travel at 12 knots with respect to the riverbank. (A knot is a nautical mile per hour, and a nautical mile is 6080.20 feet.) But there is a current in the river, and it flows at a steady 5 knots from west to east. What is the speed of your boat relative to the riverbank?

(a) 7 knots.

(b) 12 knots.

(c) 13 knots.

(d) 17 knots.

(e) More information is necessary to answer this question.

63. Suppose you have a standard 52-card deck. The deck is complete, and there is exactly one of each card (4 suits with 13 cards in each suit, and no jokers). You draw a card from this deck, then put the card back, shuffle the deck several times, and then draw another card. Then you put the card back, reshuffle the deck again several

times, and draw a third card. What is the probability that you'll get three cards of the same suit?
(a) 1/128
(b) 1/64
(c) 1/32
(d) 1/16
(e) 1/8

64. The quotient of two negative real numbers is
(a) always negative
(b) negative if the denominator is greater than −1, and positive if the denominator is less than −1
(c) negative if the denominator is less than −1, and positive if the denominator is greater than −1
(d) always positive
(e) equal to 0

65. Which, if any, of the following (a), (b), (c), or (d) cannot be defined as a scalar quantity?
(a) The mass of an apple.
(b) The temperature of the water in a swimming pool.
(c) The speed of the wind.
(d) The brightness of a light bulb.
(e) All of the above (a), (b), (c), and (d) can be defined as scalar quantities.

66. Suppose you are standing exactly 500 feet away from the base of a vertical communications tower. Your eyes are exactly 5 feet above the ground, which is flat and level in the vicinity of the tower. There is a flashing light on the top of the tower, and this light is exactly 400 feet above the ground. You look up at the light. At what elevation angle is your line of sight, relative to the horizon? Round the answer off to the nearest tenth of a degree.
(a) 38.3°
(b) 38.7°
(c) 51.3°
(d) 51.7°
(e) More information is needed to answer this question.

67. Suppose the communications tower in the scenario of Question 66 is blown out of vertical alignment by a high wind. You stand the same distance away from the base as before, that is, 500 feet. Your eyes are still exactly 5 feet above the ground. If you look at

the flashing light now, what is the angle of your line of sight relative to the horizon?

(a) 38.3°

(b) 38.7°

(c) 51.3°

(d) 51.7°

(e) More information is needed to answer this question.

68. What is the value of the quotient $(6 \times 10^5)/(2 \times 10^{-2})$?

(a) There is no way to know because the expression is ambiguous.

(b) 3×10^{-10}

(c) 3×10^3

(d) 12×10^3

(e) 3×10^7

69. In Fig. Exam-4, the independent variable is

(a) time

(b) temperature

(c) location

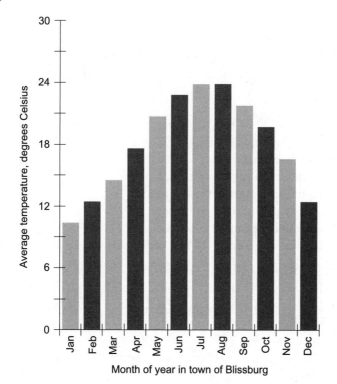

Fig. Exam-4. Illustration for Final Exam Questions 69 and 70.

(d) the absolute frequency

(e) the cumulative absolute frequency

70. In Fig. Exam-4, the dependent variable is
    (a) time
    (b) temperature
    (c) location
    (d) the absolute frequency
    (e) the cumulative absolute frequency

71. In the hexadecimal number system:
    (a) the set of possible digits is $\{0, 1\}$
    (b) the set of possible digits is $\{0, 1, 2, 3\}$
    (c) the set of possible digits is $\{0, 1, 2, 3, 4, 5, 6, 7\}$
    (d) the set of possible digits is $\{0, 1, 2, 3, 4, 5, 6, 7, 8, 9\}$
    (e) the set of possible digits is $\{0, 1, 2, 3, 4, 5, 6, 7, 8, 9, A, B, C, D, E, F\}$

72. What is 150% of 20?
    (a) 10
    (b) 20
    (c) 30
    (d) 40
    (e) It is meaningless; percentages can never be greater than 100%.

73. Let S be the sequence of all the positive whole numbers divisible by 7, considered in ascending order. That is:

$$S = 7, 14, 21, 28, 35, \ldots$$

This is an example of
    (a) a finite arithmetic sequence
    (b) an infinite arithmetic sequence
    (c) a finite geometric sequence
    (d) an infinite geometric sequence
    (e) none of the above

74. What is wrong with Fig. Exam-5?
    (a) The curve should represent a function, but it does not.
    (b) The curve should be a straight line, but it is not.
    (c) The horizontal axis does not portray a reasonable independent variable.
    (d) The vertical axis does not portray a reasonable dependent variable.
    (e) Nothing is wrong with Fig. Exam-5.

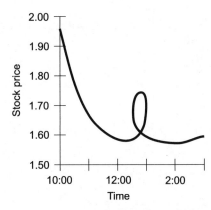

Fig. Exam-5. Illustration for Final Exam Question 74.

75. Which of the following is not a circular function?
    (a) Sine.
    (b) Cosine.
    (c) Logarithm.
    (d) Tangent.
    (e) Cosecant.

76. Consider a point (x,y,z) in 3D rectangular coordinates. The distance d of this point from the origin can be found using this formula:

$$d = (x^2 + y^2 + z^2)^{1/2}$$

Imagine the origin of a 3D rectangular coordinate system as the point where two walls intersect the ceiling in a room whose walls run north–south and east–west. A spider has built a web in the northwestern corner of this room, near the ceiling. The spider sits in its web, waiting for a fly to get caught. As the spider lurks, it is at a point 30 centimeters below the ceiling, 20 centimeters to the south of the north wall, and 40 centimeters to the east of the west wall. How far, in a straight line, is the spider from the origin? Round the answer off to the nearest centimeter.
    (a) 9 centimeters.
    (b) 29 centimeters.
    (c) 30 centimeters.
    (d) 54 centimeters.
    (e) More information is needed to answer this question.

77. Suppose, in the scenario described in Question 76, a fly gets caught in the web, 10 centimeters away from the spider. How far, measured in

a straight line, is the fly from the origin? Round the answer off to the nearest centimeter.
(a) 9 centimeters.
(b) 29 centimeters.
(c) 30 centimeters.
(d) 54 centimeters.
(e) More information is needed to answer this question.

78. If you want to design a stool so it won't wobble, even if the floor is not flat, how many legs should the stool have?
(a) 3
(b) 4
(c) 5
(d) 6
(e) 8

79. Suppose you are standing 200 yards away from a communications tower that is 150 yards high. The ground between you and the tower is flat and level. You draw this situation and get Fig. Exam-6, where d represents the distance of your shoes from the tower, e represents the height of the tower, and f represents the distance of your shoes from the top of the tower. According to the Theorem of Pythagoras, you know this:

$$d^2 + e^2 = f^2$$

You also know that there are exactly 3 feet in a yard. How far are your shoes from the top of the tower? Express the answer to the nearest 10 feet.
(a) 450 feet.
(b) 600 feet.
(c) 750 feet.

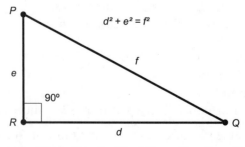

Fig. Exam-6. Illustration for Final Exam Questions 79 and 80.

(d) 900 feet.

(e) More information is necessary to answer this question.

80. Suppose, in the situation shown by Fig. Exam-6, and described by Question 79, you move twice as far away from the tower as you were before. How far are your shoes from the top of the tower now? Express the answer to the nearest 10 feet.
    (a) 1170 feet.
    (b) 1220 feet.
    (c) 1280 feet.
    (d) 1440 feet.
    (e) More information is necessary to answer this question.

81. Suppose you are grocery shopping and you come across apples on sale for $0.50 (50¢) a pound. The scale for weighing them, however, shows kilograms (kg). You place apples on the scale until it reads 3.50 kg. What is the price of this mass of apples?
    (a) 80¢
    (b) $1.59
    (c) $3.85
    (d) $7.70
    (e) It cannot be determined without more information.

82. A continuous variable
    (a) can attain only certain values over a specified range
    (b) can attain infinitely many values over a specified range
    (c) can always be portrayed as a straight line
    (d) can only be portrayed as a series of points
    (e) is a bell-shaped curve

83. Suppose a truck weighs 8459 kilograms (kg). How many grams (g) is this, written in scientific notation?
    (a) 8459 g
    (b) 8.459 g
    (c) 8.459×10^3 g
    (d) 8.459×10^6 g
    (e) 8.459×10^9 g

84. Let q represent a set of items or objects taken r at a time in a specific order. The possible number of combinations in this situation is symbolized $_qC_r$ and can be calculated as follows:

$$_qC_r = q!/[r!(q-r)!]$$

Now suppose you have a pair of standard 6-faced dice. If you toss both dice at once, how many different combinations of numbers is it possible to get?

(a) 15

(b) 30

(c) 360

(d) 720

(e) More information is necessary to answer this question.

85. Suppose you build a canvas teepee. The floor is circular with a radius of 3 meters, as shown in Fig. Exam-7. The teepee is a right circular cone supported by a central pole 4 meters high (pole not shown). The slant height is 5 meters. The surface area of a right circular cone, not including the base, is equal to π times the radius of the base, times the slant height. How much canvas will you need to construct this teepee? Neglect the small area of the exhaust fan hole at the top. Assume that the door will be a flap of canvas, that there will be no windows, and that the floor will be dirt. Let $\pi = 3.14$.

(a) 15 square meters.

(b) 28.26 square meters.

(c) 47.1 square meters.

(d) 37.68 cubic meters.

(e) 60 cubic meters.

86. The volume of a right circular cone is equal to π times the square of the radius of the base, times the height, divided by 3. How much air is

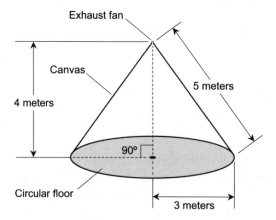

Fig. Exam-7. Illustration for Final Exam Questions 85 through 87.

enclosed by the teepee shown in Fig. Exam-7? Neglect the small volume taken up by the supporting pole. Let $\pi = 3.14$.
(a) 15 square meters.
(b) 28.26 square meters.
(c) 47.1 square meters.
(d) 37.68 cubic meters.
(e) 60 cubic meters.

87. The area of a circle is equal to the product of π and the square of its radius. How much floor area does the teepee shown in Fig. Exam-7 have? Neglect the small area taken up by the base of the supporting pole. Consider the value of π to be 3.14.
    (a) 15 square meters.
    (b) 28.26 square meters.
    (c) 47.1 square meters.
    (d) 37.68 cubic meters.
    (e) 60 cubic meters.

88. Suppose you graph two different equations in two variables. Both of the equations have graphs that look like circles. How many common solutions are there to this pair of equations?
    (a) There are no solutions.
    (b) There is one solution.
    (c) There is more than one solution.
    (d) There are infinitely many solutions.
    (e) More information is needed to answer this.

89. Imagine that you are a dedicated lap swimmer, and you are used to swimming 3000 meters every day, which is 30 round-trip laps (a lap is two one-way lengths) in your local 50-meter pool. You are in a strange town and you go to their local pool, and discover that it is 25 yards long. You happen to remember that there are approximately 39.37 inches in 1 meter, and that 1 yard is exactly 36 inches. Based on this information, how many round-trip laps of the 25-yard pool should you swim in order to be sure you get your 3000-meter workout, but without swimming any more laps than necessary?
    (a) More information is necessary in order to answer this question.
    (b) 28
    (c) 33
    (d) 55
    (e) 66

90. The median for a discrete variable in a statistical distribution is
    (a) the smallest value
    (b) the largest value
    (c) the mathematical average of all the values
    (d) the value in the exact middle
    (e) the value that occurs most often

91. In the octal number system:
    (a) the set of possible digits is {0, 1}
    (b) the set of possible digits is {0, 1, 2, 3}
    (c) the set of possible digits is {0, 1, 2, 3, 4, 5, 6, 7}
    (d) the set of possible digits is {0, 1, 2, 3, 4, 5, 6, 7, 8, 9}
    (e) the set of possible digits is {0, 1, 2, 3, 4, 5, 6, 7, 8, 9, A, B, C, D, E, F}

92. In a major-league baseball diamond, the distance between home
    plate and first base is 90 feet. The same is true of the distance between
    first base and second base, between second base and third base, and
    between third base and home plate. The pitcher's rubber is 60.5 feet
    from home plate. Suppose a triangle is laid out with straight ropes
    between home plate, first base, and the pitcher's rubber. What is
    the sum of the measures of the interior angles of this triangle?
    (a) 172°
    (b) 176.5°
    (c) 180°
    (d) 183.5°
    (e) More information is needed to answer this question.

93. Suppose that major-league baseball officials decide to change the
    dimensions of a standard baseball diamond. The motivation is to
    make it more difficult for pitchers, causing higher-scoring games.
    According to the new regulations, the pitcher's rubber will be 70
    feet, rather than 60.5 feet, away from home plate. Further imagine
    that the distances between the bases will be left the same, at 90
    feet. Suppose a triangle is laid out with straight ropes between
    home plate, first base, and the pitcher's rubber in a diamond with
    the new official dimensions. What is the sum of the measures of
    the interior angles of this triangle?
    (a) 172°
    (b) 176.5°
    (c) 180°
    (d) 183.5°
    (e) More information is needed to answer this question.

94. What is the sum of 2/3 and 4/5?
    (a) 6/15
    (b) 8/15
    (c) 3/4
    (d) 1
    (e) 22/15

95. Consider the following equation:

$$(x - 7)(x + 2) = 0$$

What is the real-number solution set for this equation?
(a) $\varnothing$ (The empty set.)
(b) $\{-7\}$
(c) $\{2\}$
(d) $\{-7, 2\}$
(e) $\{7, -2\}$

96. A kilogram (kg) represents 1000 grams, and a microgram (µg) represents 0.000001 grams. Suppose a sand grain has a mass of 2 µg while a bag of potatoes has a mass of 2 kg. How many orders of magnitude more massive is the bag of potatoes than the sand grain?
(a) More information is needed to answer this.
(b) 2
(c) 3
(d) 9
(e) 1,000,000,000

97. Suppose you buy stock in Corporation X on December 31 of a certain year and then forget all about it. One day in the middle of January you get a report from the company saying that your stock has increased in value by $5 a share each year for the past 10 years. Every December 31, the stock has been worth $5 a share more than it was on December 31 of the previous year. If you graph the stock price per share based on 11 points (the initial price, and the price on December 31 for each of the 10 years following) and then connect the points with a smooth curve, what will the curve look like? Assume the horizontal graph scale shows time, and the vertical scale shows the stock price per share, and both of the scales are linear.
(a) The curve will be a parabola.
(b) The curve will be a circle.
(c) The curve will be a straight line.

(d) You won't be able to connect the points with a smooth curve, because they'll be scattered all over the place.
(e) More information is necessary to answer this.

98. Let q represent a set of items or objects taken r at a time in a specific order. The possible number of permutations in this situation is symbolized $_qP_r$ and can be calculated as follows:

$$_qP_r = q!/(q-r)!$$

Now suppose you have a pair of standard 6-faced dice. If you toss both dice at once, how many different permutations of numbers is it possible to get?
(a) 15
(b) 30
(c) 360
(d) 720
(e) More information is necessary to answer this question.

99. Figure Exam-8 represents
(a) vector addition in 2 dimensions
(b) vector addition in 3 dimensions
(c) scalar addition in 2 dimensions
(d) scalar addition in 3 dimensions
(e) none of the above

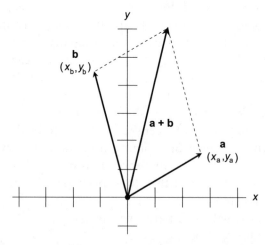

Fig. Exam-8. Illustration for Final Exam Questions 99 and 100.

100. In Fig. Exam-8, let each horizontal division represent 10 units, and let each vertical division represent 5 units. Suppose $\mathbf{a} = (28,7)$ and $\mathbf{a} + \mathbf{b} = (13,30)$. What is $\mathbf{a} + (\mathbf{a} + \mathbf{b})$?

(a) 574

(b) 78

(c) (58,20)

(d) (41,37)

(e) More information is needed to answer this question.

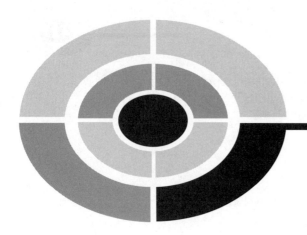

Answers to Quiz, Test, and Exam Questions

Chapter 1

1. c	2. a	3. c	4. c	5. b
6. a	7. b	8. d	9. a	10. b

Chapter 2

1. d	2. d	3. a	4. a	5. c
6. d	7. a	8. b	9. b	10. c

Chapter 3

1. d	2. a	3. c	4. b	5. d
6. a	7. c	8. c	9. d	10. b

Answers

Chapter 4

1. d	2. b	3. b	4. d	5. d
6. c	7. a	8. c	9. d	10. b

Test: Part 1

1. d	2. b	3. a	4. b	5. c
6. e	7. a	8. b	9. b	10. c
11. b	12. d	13. a	14. e	15. d
16. a	17. b	18. a	19. b	20. c
21. c	22. a	23. c	24. a	25. c
26. a	27. e	28. e	29. a	30. e
31. e	32. c	33. d	34. c	35. e
36. a	37. b	38. b	39. d	40. e

Chapter 5

1. b	2. a	3. c	4. a	5. c
6. d	7. a	8. b	9. a	10. d

Chapter 6

1. a	2. a	3. d	4. d	5. d
6. c	7. c	8. d	9. c	10. b

Chapter 7

1. c	2. a	3. c	4. a	5. b
6. b	7. d	8. a	9. a	10. b

Chapter 8

1. a	2. c	3. d	4. a	5. c
6. a	7. d	8. c	9. b	10. b

Test: Part 2

1. a	2. d	3. c	4. e	5. b
6. a	7. e	8. c	9. b	10. c
11. d	12. d	13. d	14. e	15. c
16. d	17. b	18. c	19. e	20. e
21. b	22. a	23. b	24. c	25. c
26. c	27. d	28. c	29. c	30. b
31. b	32. a	33. e	34. a	35. d
36. a	37. a	38. c	39. a	40. b

Chapter 9

1. c	2. b	3. d	4. c	5. c
6. c	7. c	8. b	9. d	10. a

Answers

Chapter 10

1. b	2. a	3. a	4. c	5. a
6. d	7. c	8. a	9. d	10. b

Chapter 11

1. c	2. d	3. c	4. d	5. b
6. b	7. c	8. d	9. b	10. a

Chapter 12

1. a	2. b	3. b	4. a	5. c
6. b	7. d	8. d	9. b	10. d

Test: Part 3

1. a	2. e	3. c	4. c	5. d
6. d	7. b	8. d	9. e	10. c
11. c	12. e	13. c	14. a	15. d
16. e	17. e	18. d	19. d	20. a
21. a	22. e	23. d	24. e	25. c
26. a	27. a	28. c	29. c	30. c
31. a	32. e	33. c	34. a	35. d
36. a	37. d	38. b	39. e	40. c

Chapter 13

1. c	2. a	3. a	4. b	5. b
6. d	7. c	8. d	9. d	10. a

Chapter 14

1. d	2. b	3. c	4. d	5. b
6. b	7. b	8. b	9. a	10. a

Chapter 15

1. d	2. d	3. a	4. d	5. b
6. a	7. a	8. b	9. d	10. a

Test: Part 4

1. b	2. c	3. c	4. b	5. d
6. b	7. e	8. d	9. b	10. d
11. c	12. d	13. d	14. c	15. d
16. c	17. d	18. d	19. b	20. e
21. a	22. b	23. c	24. b	25. c
26. b	27. e	28. a	29. a	30. c
31. b	32. e	33. a	34. c	35. d
36. b	37. d	38. c	39. c	40. a

Final Exam

1. d	2. b	3. a	4. d	5. c
6. a	7. c	8. c	9. a	10. c
11. e	12. c	13. c	14. e	15. d
16. d	17. b	18. e	19. b	20. b
21. a	22. d	23. a	24. d	25. c
26. b	27. b	28. a	29. d	30. c
31. a	32. c	33. a	34. b	35. e
36. c	37. b	38. b	39. d	40. b
41. d	42. d	43. d	44. c	45. e
46. a	47. c	48. d	49. d	50. d
51. e	52. d	53. a	54. e	55. d
56. c	57. d	58. d	59. b	60. a
61. d	62. c	63. b	64. d	65. e
66. a	67. e	68. e	69. a	70. b
71. e	72. c	73. b	74. a	75. c
76. d	77. e	78. a	79. c	80. c
81. c	82. b	83. d	84. a	85. c
86. d	87. b	88. e	89. e	90. d
91. c	92. c	93. c	94. e	95. e
96. d	97. c	98. b	99. a	100. d

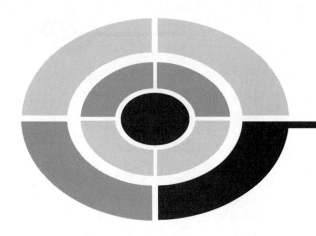

Suggested Additional References

Books

Gibilisco, Stan, *Geometry Demystified*. McGraw-Hill, New York, NY, 2003.

Gibilisco, Stan, *Statistics Demystified*. McGraw-Hill, New York, NY, 2004.

Gibilisco, Stan, *Trigonometry Demystified*. McGraw-Hill, New York, NY, 2003.

Huettenmueller, Rhonda, *Algebra Demystified*. McGraw-Hill, New York, NY, 2003.

Krantz, Steven, *Calculus Demystified*. McGraw-Hill, New York, NY, 2003.

Prindle, Anthony and Katie, *Math the Easy Way – 3rd Edition*. Barron's Educational Series, Hauppauge, NY, 1996.

Webber, John, *Math for Business and Life – 2nd Edition*. Olympus Publishing Company, Salt Lake City, UT, 2003.

Web Sites

Encyclopedia Britannica Online, www.britannica.com
Eric Weisstein's World of Mathematics, www.mathworld.wolfram.com

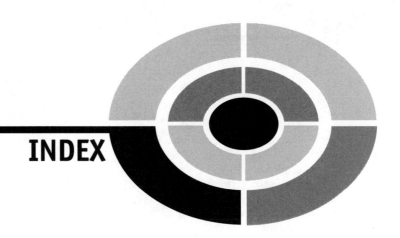

INDEX

abscissa, 252
absolute zero, 79
acceleration, 373–8
acceleration meter, 375
acceleration of gravity, 375–6
accelerometer, 375
acute angles, 207–8
additive identity element, 19
additive inverses, 20
adjacent sides, 288
aleph-one, 17
aleph-nought, 17
aleph-null, 17
alternate exterior angles, 205–6
alternate interior angles, 205
amount of matter, 75
ampere, 75, 79
amplitude of sine wave, 280
angle-angle-angle, 210
angle bisection, 201–2
angle-side-angle, 209–10
angles
  acute, 207–8
  addition of, 203–4
  alternate exterior, 205–6
  alternate interior, 205
  corresponding, 206
  obtuse, 207–8
  notation for, 201
  subtraction of, 203–4
  vertical, 204
angular degree, 83–4, 276

angular measure
  plane, 83–4
  solid, 84
apex, 235–6
arc minute, 276
arc second, 276
arctangent, 311
arithmetic, 24
arithmetic extrapolation, 334–8
arithmetic interpolation, 334–8
arithmetic progression, 333, 335–6
arithmetic sequence, 333, 335–6
arithmetic series, 333–4
associative law
  for addition, 21
  for multiplication, 21
average speed, 369
Avogadro's number, 80
azimuth, 368, 373

base 2, 8, 9, 345–6
bar graph
  horizontal, 39–40, 42–3
  vertical, 39
base units, 75–83
bimodal distribution, 160
binary number system, 9
blackbody, 80
brightness of light, 75

candela, 75, 80
candlepower, 80

Cantor, Georg, 17
cardinals, transfinite, 17
cardioid, 268–70
Cartesian plane, 250–60, 309–13
Cartesian three-space, 309, 313–29
census, 147
centimeter, 75
centimeter/gram/second system, 75
circle, 224, 263–4, 274–7
circular functions
    primary, 277–83
    secondary, 283–7
circular motion, 279
circumference, 224
clinometer, 373
coefficients, 119
combinations, 183–4
common exponential, 346
commutative law
    for addition, 20
    for multiplication, 20–1
compass, points of, 373
complementary outcomes, 175–6
complex numbers, 129–32
cone, 240–3
constant function, 48
continuous variable, 145–6
Continuum Hypothesis, 16
convergent series, 341
conversions, 87–91
coordinate plane
    Cartesian, 250–60
    log-log, 357–8
    Polar, 260–71
    rectangular, 250–60
    semilog, 356–7
correlation, 47–8
corresponding angles, 206
cosecant, 283
cosecant function, 283–4
cosine, 264–5, 281
cosine function, 281–2
cosine wave, 281–2
cotangent, 285
cotangent function, 285–6
coulomb, 85–6
counting numbers, 8
cross multiplication, 25

cross product, 320–1
cube, 237
cubic equation, 133–5
cubic formula, 133
cumulative absolute frequency, 154–5
cumulative relative frequency, 155–6
curve fitting, 44–5
cycle, 86–7
cycle per second, 86–7
cylinder, 241, 243–4

decibel, 354–5
decimal number system, 8
decimal
    nonterminating, nonrepeating, 16
    nonterminating, repeating, 15
    terminating, 14
decimal point, 54
degree, angular, 83–4, 276
degree Celsius, 75
degree Fahrenheit, 75
degree of arc, 276
dependent variable, 34, 252, 315
derived units, 83
difference, 11
dimension, 88
dimensionless unit, 80
direction of vector, 309, 311
direction angle, 277–8, 318–20, 368
direction cosine, 318–319
direction numbers, 327
discrete values, 38
discrete variable, 144–5
displacement, 75, 76, 367–8
distance
    addition of, 203
    between points, 252–3, 315–16
    subtraction of, 203
    notation for, 200
    versus displacement, 367
distribution, 39–40, 149–54, 160
distributive laws, 21
divergent series, 340
dot product, 312–13, 320

electric charge quantity, 85
electric current, 75, 79

electric potential, 86
electrical resistance, 86
electromotive force, 86
element, 146
elevation, 373
ellipse, 224–5, 264–5
empirical evidence, 170
empirical probability, 170–2
energy, 85
English system, 75
equilateral triangle, 212
equation
  cubic, 133–5
  linear, single-variable, 105–10
  linear, two-by-two, 110–18
  one-variable, nth-order, 138–140
  one-variable, higher-order, 133–40
  one-variable, second-order, 119–29
  quadratic, 119–29
  quartic, 135–6
  quintic, 136–8
equivalent vectors, 310, 316–17
error
  instrument, 146
  observation, 146
Euclidean geometry, 206
event, 147, 167
experiment, 143–4
exponential function, 345–50
exponentials
  common, 346
  exponential of, 349
  natural, 346–7
  product of, 348
  ratio of, 348
  reciprocal of, 348
exponential function, 345–50
exponents, plain text, 55–6
extrapolation
  arithmetic, 334–8
  geometric, 342
  linear, 44–6
extreme numbers, 52–73

facets, 235
factored form of quadratic equation,
  120–1
factorial, 181–3

foot, 75
foot/pound/second system, 75
force, 362–7, 376
force meter, 367
four-leafed rose, 267, 269
frequency
  absolute, 148
  cumulative absolute, 154–5
  cumulative relative, 155–6
  of oscillation, 364
  of signal wave, 86–7
  of sine wave, 280
  relative, 148
frequency distribution, 151–4
  grouped, 153–4
  ungrouped, 152
function
  constant, 48, 262–3
  cosecant, 283–4
  cosine, 281–2
  cotangent, 285–6
  definition of, 36–7, 262–3
  exponential, 345–50
  logarithmic, 352–9
  nondecreasing, 47
  nonincreasing, 46–7
  secant, 284–5
  sine, 278–80
  tangent, 281–3
  trending downward, 47
  trending upward, 47
Fundamental Theorem of
  Arithmetic, 13
fuzzy truth, 166

gamma function, 181
geometric interpolation, 341–2
geometric progression, 339, 342–5
geometric sequence, 339, 342–5
geometric series, 340–1
geometry
  Euclidean, 206
  non-Euclidean, 206
  plane, 199–228
  solid, 229–49
gigahertz, 87
Global Positioning System, 368
gram, 75

graphs, 38–43
  histogram, 39–40
  horizontal bar, 39–40, 42–3
  nomograph, 40–1
  point-to-point, 41–2
  resolution of, 42
  scales in, 42
  vertical bar, 39

half plane, 232
hertz, 86–7
hexadecimal numbers, 9
histogram, 39–40
horizontal bar graph, 39–40, 42–3
hundredths digit, 8
hyperbola, 265–6
hyperspace, 230
hypotenuse, 212, 252, 288

identity element
  additive, 19
  multiplicative, 19–20
imaginary number, 129–30
imaginary number line, 129–30
impulse, 378–81
increment, 144
independent outcomes, 173
independent variable, 34, 252, 315
inflection point, 135, 138
instantaneous speed, 369
integer roots, 27
integers
  addition of, 11
  definition of, 10
  division of, 11–12
  exponentiation with, 12
  multiplication of, 11
interior angles
  of plane quadrilateral, 219
  of triangle, 210–11
interior area
  of circle, 224
  of parallelogram, 220
  of rectangle, 221
  of rhombus, 220
  of square, 221
  of trapezoid, 222
  of triangle, 213–14

International System, 75, 80–1
interpolation
  arithmetic, 334–8
  geometric, 341–2
  linear, 43–4, 48–9
intersecting line principle, 231
intersecting planes, 233
inverse tangent, 311
inverses
  additive, 20
  multiplicative, 20
irrational numbers, 16
irregular pyramid, 236
isosceles triangle, 211

joule, 85

Kelvin, 75, 79
kilogram, 75, 77
kilohertz, 87
kilohm, 86
kilometer, 77
kilometers per hour, 369
kilomole, 80
kilovolt, 86
kilowatt, 85

large numbers, law of, 172–3
lateral surface area, 242, 244
law of large numbers, 172–3
lemniscate, 266–7
light-second, 78
line
  definition of, 229–30
  direction numbers, 327
  in three-space, 326–9
  parametric equations for, 327–8
  point-slope form of, 257–8
  slope-intercept form of, 254–6
  symmetric-form equation for, 327
line and point principle, 231
linear axis, 356
linear equations
  addition method for solving, 112–13
  basic, in one variable, 106
  graphical method for solving, 113–16
  illustrating the solution of, 107–10
  single-variable, 105–18
  standard form, one variable, 106–7

substitution method for solving, 111
two-by-two, 110–16
linear extrapolation, 44–6
linear interpolation, 43–4, 48–9
linear scale, 356
liter, 239–40
logarithmic axis, 356
logarithmic function, 352–9
logarithmic scale, 356
logarithms
    common, 350–1
    graphs based on, 355–9
    Napierian, 351–2
    natural, 351–2
    of product, 352
    of quantity raised to a power, 353
    of ratio, 353
    of reciprocal, 353
    of root, 353–4
log-log coordinates, 357–8
luminous intensity, 80

magnitude of vector, 309–11, 317–18
major diagonal, 214
major half-axis, 224
major semi-axis, 265
mass, 75, 77, 362–7, 376
mass meter, 363–6
material quantity, 80
mathematical probability, 169–70
mean
    arithmetic, 24
    in statistics, 156–8
mean solar day, 78
mechanical force, 84
median
    in statistics and probability, 159–60
    of trapezoid, 218–19
megahertz, 87
megavolt, 86
megawatt, 85
megohm, 86
meter, 75, 76
meter stick, 368
meter/kilogram/second system, 75
meters per second, 369
meters per second per second, 374–5
meters per second squared, 374–5

metric system, 75
metric ton, 77
microampere, 79
microgram, 77
micrometer, 76–7
micron, 77
microsecond, 79
microvolt, 86
microwatt, 85
midpoint principle, 201
miles per hour, 369
milliampere, 79
milligram, 77
millimeter, 76
millimole, 80
millisecond, 79
millivolt, 86
milliwatt, 85
minor diagonal, 214
minor half-axis, 224
minor semi-axis, 265
minute of arc, 276
mitosis, 344
mode, 160–1
mole, 75, 80
momentum, 378–81
multiple outcomes, 178–81
multiplication by zero, 24
multiplicative identity element, 19–20
multiplicative inverses, 20
mutual perpendicularity, 206–7
mutually exclusive outcomes, 174–5

nanoampere, 79
nanogram, 77
nanometer, 77
nanosecond, 79
nanowatt, 85
natural exponential, 346–7
natural logarithm base, 181
natural numbers, 8
negative radius, 261–2
newton, 79, 84–5, 366–7
nomograph, 40–1
nondecreasing function, 47
nondisjoint outcomes, 176–8
non-Euclidean geometry, 206
nonincreasing function, 46–7

nonstandard directions, 262
nonterminating, nonrepeating decimal, 16
nonterminating, repeating decimal, 15
normal vector, 324
nth-order equation, 138–40
numbering systems, 8–10
numbers
  binary, 9
  decimal, 8
  counting, 8
  extreme, 52–73
  hexadecimal, 9
  irrational, 16
  natural, 8
  octal, 9
  operations with, 19–23
  prime, 12–13
  rational, 14–16
  real, 16–17
  whole, 8
numerals, 8

obtuse angles, 207–8
octal numbers, 9
odometer, 368
ohm, 86
ones digit, 8
one-variable, higher-order equation, 133–40
one-variable, second-order equation,
    119–29
ordered pair, 251–2
ordered triple, 314–15
order of magnitude, 56–8, 349, 356–9
ordinate, 252
orientation of vector, 309
origin, 11, 252, 314
oscillation period, 364
oscillation frequency, 364
outcomes
  complementary, 175–6
  definition of, 167
  independent, 173
  multiple, 178–81
  mutually exclusive, 174–5
  nondisjoint, 176–8

parabola, 124–9
parallel lines, 233

parallel principle, 206–7
parallelepiped, 238–9
parallelogram, 215–16, 219–20
parameter
  of population, 148
  of straight line, 328
parametric equations, 327–8
partial sums, 340–1
perimeter
  of circle, 224
  of parallelogram, 219–20
  of rectangle, 221
  of square, 221
  of trapezoid, 222
  of triangle, 213
period, of oscillation, 364
permutations, 183
perpendicular bisector, 202–3
perpendicularity, 202, 206–7
pi, 17
plain-text exponents, 55–6
plane
  angular measure, 83
  equation of, in three-space, 323–4
  geometry, 199–228
  in space, 323–6
  plotting, in three-space, 324–6
  region, 231–2
point, 229–30
point-point-point, 208
point-slope form, 257–8
point-to-point graph, 41–2
points of the compass, 373
polar coordinate plane, 260–71
polyhedron, 235
population, 146–9
population mean, 156
potential difference, 86
pound, 75
power, 85
power-of-10 notation, 54–67
  addition in, 60–1
  division in, 62–3
  exponentiation in, 62–3
  in calculators, 183
  multiplication in, 61–2
  subtraction in, 61
  taking roots in, 64

powers
  definition of, 12
  difference of, 28
  negative, 28
  of difference, 29
  of reciprocal, 29
  of signs, 25
  of sum, 29
  product of, 28
  quotient of, 28
  rational-number, 27
  sum of, 28
precedence of operations, 22, 66–7
prefix multipliers, 58–9
primary circular functions, 277–83
prime factors, 13
prime numbers, 12–13
principle of n points, 200
prism, rectangular, 238
probability, 165–86
  empirical, 170–2
  mathematical, 169–70
probability fallacy, 165–7
product
  definition of, 11
  of quotients, 26
  of signs, 24
  of sums, 25
progression
  arithmetic, 333, 335–6
  geometric, 339, 342–5
proportion, 12
pyramid, 236–7
Pythagorean Theorem, 212–13, 252–3,
    288–92, 311

quadratic equation, 119–29
  standard form of, 119
  factored form of, 120–1
quadratic formula, 123–4
quadrilaterals, 214–23
  interior angles of, 219
quantum mechanics, 166
quartic equation, 135–6
quintic equation, 136–8
quotient
  definition of, 11–12
  of products, 26

  of quotients, 26
  of signs, 24

radian, 83, 260–1
radius, 260–2
radix 10, 8
radix point, 8, 54
random sample, 147
random variable, 147
relation, 35–6, 262–3
ratio, 12, 14
rational numbers, 14–16
real numbers, 16–17
reciprocal
  of product, 25
  of quotient, 26
  of reciprocal, 25
rectangle, 216–17, 221
rectangular 3D coordinates, 313–29
rectangular coordinate plane, 250–60
rectangular prism, 238
rectangular three-space, 309
reference frame, 379
regular hexahedron, 237
regular pyramid, 236
regular tetrahedron, 235
rhombus, 217–18, 220
right circular cone, 241–2
right circular cylinder, 243–4
right-hand rule, 320
right triangle model, 287–90
rounding, 67

sample, 146–9
sample mean, 156
sample space, 167–9
scalar product, 312, 320
scalar quantity, 363,369
scatter plots, 315
scientific notation, 54–67
  addition in, 60–1
  division in, 62–3
  exponentiation in, 62–3
  in calculators, 183
  multiplication in, 61–2
  subtraction in, 61
  taking roots in, 64
secant, 284

secant function, 284–5
second, 75, 78–9
second of arc, 276
secondary circular functions, 283–7
semi-axes, 265
semilog coordinates, 356–7
sequence
  arithmetic, 333, 335–6
  geometric, 339, 342–5
series
  arithmetic, 333–4
  convergent, 341
  divergent, 340
  geometric, 340–1
  partial sums in, 340–1
sets
  coincident, 7
  disjoint, 7
  intersection, 6
  union, 6
  proper, 7
  subsets, 7
side-angle-side, 209
side-side-side, 209
significant digits, 68
significant figures, 68–71
simple plane region, 231–2
sine, 239, 264–5, 278–9
sine function, 278–80
sine wave, 280
sinusoid, 280
skew lines, 233–4
slant circular cone, 242–3
slant circular cylinder, 244
slant height, 236
slope, 255, 259–60
slope-intercept form, 254–6
solar day, 78
solid angular measure, 84
solid geometry, 229–249
space, 229–30
speed, 368–71
speedometer, 369
sphere, 245
spiral, 267, 270
spiral of Archimedes, 267
square, 221
statistic of sample, 148–9

statistics, 143–64
steradian, 80, 84
subscripts, 52–3
superscripts, 52–3
sum
  definition of, 11
  of quotients, 27
surface area
  of cube, 237
  of parallelepiped, 239
  of pyramid, 236
  of rectangular prism, 238
  of regular tetrahedron, 235
  of right circular cone, 241–2
  of right circular cylinder, 243–4
  of sphere, 245
symbology, 81
symmetric-form equation, 327
systems of units, 75

tachometer, 369
tangent, 281
tangent function, 281–3
temperature, 75, 79
tens digit, 8
tenths digit, 9
terahertz, 87
terminating decimal, 14
tetrahedron, 235–6
Theorem of Pythagoras, 212–13, 252–3,
  288–92
three-leafed rose, 166–8
three point principle, 200, 230–1
time, 75, 78–9
"times sign," 55
transfinite cardinals, 17
trapezoid, 218–19, 222
trending downward, 47
trending upward, 47
triangles, 208–214
  congruent, 209
  equilateral, 212
  interior area of, 213–14
  isosceles, 211
  perimeter of, 213
  similar, 210
triangulation, 368
trigonometry, 274–94

trimodal distribution, 160
triple point, 79
truncation, 66
two point principle, 200

unit circle, 275, 277
unit imaginary number, 129
unit vectors, 321–2
units, 74–92
unknown, 144

variable
  continuous, 145–6
  in algebra, 33–51
  dependent, 34, 252
  discrete, 144–5
  in Cartesian three-space, 15
  in statistics, 144–6
  independent, 34, 252
  random, 147
  reversing, 37
vectors
  acceleration, 373–8
  cross product of, 320–1
  definition of, 309
  direction of, 309, 311, 318–20
  displacement, 367–8
  dot product of, 312–13, 320
  equivalent, 310, 316–17
  force, 366
  impulse, 378–81
  in Cartesian plane, 309–13
  in Cartesian three-space, 316–22
  in trigonometry, 278
  magnitude of, 309–11, 317–18
  momentum, 378–81
  multiplication by scalar, 311–12, 319–20
  normal, 324
  orientation of, 309

scalar product, 312
standard form of, 310, 316–17
sum of, 311, 319
unit, 321–2
velocity, 371–3
zero, 312
velocity, 371–3
vertex of angle, 287
vertical angles, 204
vertical bar graph, 39
vertices of triangle, 287
volt, 86
volume
  of cube, 237
  of parallelepiped, 239
  of pyramid, 236–7
  of rectangular prism, 238
  of regular tetrahedron, 236
  of right circular cone, 241–2
  of right circular cylinder, 243–4
  of slant circular cone, 242–3
  of slant circular cylinder, 244
  of sphere, 245

watt, 85
weight, 77, 363
whole numbers, 8

x axis, 250–1, 314
xyz-space, 309, 313–29

y axis, 250–1, 314
y-intercept, 255, 259–60

z axis, 314
zero denominator, 23
zero numerator, 23
zero vector, 312
zeroth power, 24

ABOUT THE AUTHOR

Stan Gibilisco is one of McGraw-Hill's most prolific and popular authors. His clear, reader-friendly writing style makes his electronics books accessible to a wide audience, and his background in mathematics and research makes him an ideal editor for professional handbooks. He is the author of the *TAB Encyclopedia of Electronics for Technicians and Hobbyists, Teach Yourself Electricity and Electronics*, and *The Illustrated Dictionary of Electronics*. *Booklist* named his *McGraw-Hill Encyclopedia of Personal Computing* a "Best Reference" of 1996.